Modeling Approaches in Avian Conservation and the Role of Field Biologists

ORNITHOLOGICAL MONOGRAPHS

Editor: John Faaborg

224 Tucker Hall
Division of Biological Sciences
University of Missouri
Columbia, Missouri 65211

Project Manager: Mark C. Penrose
Managing Editor: Richard D. Earles

AOU Publications Office
622 Science Engineering
Department of Biological Sciences
University of Arkansas
Fayetteville, Arkansas 72701

The *Ornithological Monographs* series, published by the American Ornithologists' Union, has been established for major papers and presentations too long for inclusion in the Union's journal, *The Auk*. All material in this monograph may be copied for noncommercial purposes of educational or scientific advancement without need to seek permission.

Copies of *Ornithological Monographs* are available from Buteo Books, 3130 Laurel Road, Shipman, VA 22971. Price of *Ornithological Monographs* no. 59: $10.00 ($9.00 for AOU members). Add $4.00 for handling and shipping charges in U.S., and $5.00 for shipping to other countries. Make checks payable to Buteo Books.

Authors of this issue: Steven R. Beissinger, Jeffrey R. Walters, Donald G. Catanzaro, Kimberly G. Smith, John B. Dunning, Jr., Susan M. Haig, Barry R. Noon, and Bradley M. Stith.

Library of Congress Control Number 2005937805

Printed by Cadmus Communications, Ephrata, PA 17522

Issued 31 January 2006

Ornithological Monographs, No. 59 viii + 56 pp.

ISBN: 0-943610-66-4

Cover: A field ornithologist at work, illustrating the competing needs of hands-on data collection and abstract modeling that ornithologists constantly juggle when doing their research. Original drawing by Monica Jane Albe.

MODELING APPROACHES IN AVIAN CONSERVATION AND THE ROLE OF FIELD BIOLOGISTS

BY

STEVEN R. BEISSINGER,[1] JEFFREY R. WALTERS,[2] DONALD G. CATANZARO,[3] KIMBERLY G. SMITH,[3] JOHN B. DUNNING, JR.,[4] SUSAN M. HAIG,[5] BARRY R. NOON,[6] AND BRADLEY M. STITH[7]

[1]*Department of Environmental Science, Policy and Management, University of California, Berkeley, California 94720, USA;*
[2]*Department of Biology, Virginia Polytechnic Institute and State University, Blacksburg, Virginia 24061, USA;*
[3]*Center for Advanced Spatial Technologies and Department of Biological Sciences, University of Arkansas, Fayetteville, Arkansas 72701, USA;*
[4]*Department of Forestry and Natural Resources, Purdue University, West Lafayette, Indiana 47907, USA;*
[5]*U.S. Geological Survey Forest and Rangeland Ecosystem Science Center, 3200 SW Jefferson Way, Corvallis, Oregon 97331, USA;*
[6]*Department of Fishery and Wildlife Biology, Colorado State University, Fort Collins, Colorado 80523, USA; and*
[7]*Department of Wildlife Ecology and Conservation, University of Florida, Gainesville, Florida 32611, USA*

ORNITHOLOGICAL MONOGRAPHS NO. 59

PUBLISHED BY
THE AMERICAN ORNITHOLOGISTS' UNION
WASHINGTON, D.C.
2006

TABLE OF CONTENTS

Lists of Tables and Figures vi
From the Editor vii
ABSTRACT 1
INTRODUCTION 2
THE FORM OF ECOLOGICAL MODELS 3
DETERMINISTIC SINGLE-POPULATION MATRIX MODELS 4
Structure of Matrix Population Models 5
Use of Matrix Population Models in Conservation 8
STOCHASTIC POPULATION VIABILITY ANALYSIS MODELS FOR SINGLE POPULATIONS 10
Structure of Stochastic Population Viability Analysis Models for Single Populations 11
Conservation Use of Stochastic Population Viability Analysis Models for Single Populations 12
METAPOPULATION MODELS 15
Structure of Metapopulation Models 15
Use of Metapopulation Models in Conservation 19
SPATIALLY EXPLICIT MODELS 20
Structure of Spatially Explicit Models 20
Use of Spatially Explicit Models in Conservation 23
GENETIC MODELS 26
Structure of Genetic Models 26
Effective population size 26
Pedigree analyses 27
Use of Genetic Models in Conservation 30
SPECIES DISTRIBUTION MODELS 31
Structure of Species Distribution Models 32
Use of Species Distribution Models in Conservation 35
INTELLIGENT USE OF MODELS TO MAKE CONSERVATION DECISIONS 37
Attributes of Useful Models 37
Working with Multiple Models in a Decision-theoretic Framework 39
HOW FIELD BIOLOGISTS CAN INTERACT WITH MODELERS TO IMPROVE CONSERVATION DECISIONS 40
ACKNOWLEDGMENTS 41
LITERATURE CITED 42

LIST OF TABLES

1. Structure of the standard input file ("Me, Ma, Pa" file) used to keep studbooks, to construct pedigrees, and for gene-drop analyses 29
2. Consideration of genetic factors in common population viability models used in conservation planning 30

LIST OF FIGURES

1. Matrix population models for prebreeding and postbreeding life cycles, and elasticity calculations 6
2. Simplified example of the structure and outcomes from a stochastic single-population PVA model 11
3. Demographic PVA models with various degrees of spatial explicitness 17
4. Gene-drop pedigree model and analysis for a small population 28
5. Example of a gap modeling process for the Ovenbird (*Seiurus aurocapilla*) in Arkansas 33

From the Editor

Like many members of the American Ornithologists' Union, I developed a passion for birds early in life and have been trying to make a living from this passion ever since. As you get to know fellow AOU members, particularly over drinks, you hear stories about that moment in their lives when they discovered this incredible feeling about birds that could not be ignored. Some of us started at age six or seven, others in high school or college.

Some of us, lacking the artistic skills of David Sibley or the humor of Pete Dunne, have had to fall back on doing research. Our field work allows us to earn a paycheck in pursuit of our passion. We get to see and catch and count birds in what are sometimes exciting, and hopefully always interesting, places. We are paid to write reports, scientific papers, and books that present our results from the field. In many cases, our work has direct or indirect implications for bird conservation.

Unfortunately, we cannot simply write papers about anything that catches our fancy. Rather, we need to justify our work on the basis of prior studies and present our results using up-to-date methodologies. For many of us, this is where the ugly concepts of hypotheses, models, and statistics come into play and turn our lives of passion into actual work and sometimes drudgery. During my career, the profession has become more and more based on the testing of models, usually with associated quantitative measures. I took a single statistics course during eight years of college; my graduate students take at least three or four, and often more. One even has a Master's degree in statistics to go along with her MS in ecology. More and more, our field work is only as good as the model being tested, and the test only as good as the calculations provided.

Of course, doing field work requires a lot of skill, though usually of the "natural history" category that involves details about birds, plants, and so forth. Doing modeling and developing statistical methods obviously requires a high comfort level with conceptual thinking and mathematics. Few ornithologists are experts in both areas; field people are always trying to figure out what is currently the best way to measure their area of specialty, while modelers are trying to provide methods for the analytical questions posed by the field people. The goal of both groups is to develop the best science possible, with the appropriate scientific models tested with the best quantitative techniques.

All these areas within ornithological research change over time, so it is difficult for field people to stay current with the latest models and for modelers to ensure that field people are providing the best data possible and analyzing it properly. *Ornithological Monograph No. 59* addresses this problem by serving as a bridge between field biologists and modelers. It provides a state-of-the-art review by a set of experts of the models they consider most relevant to current avian conservation. It explains why these models are relevant and then shows how they can be quantitatively tested with field data. In addition, it promotes interaction between field workers and modelers, because models are of little use if they are not tested and supported with the right data from the field. This monograph should be of great value to beginning graduate students who are planning field studies and handy for us older folks who may need to catch up on things.

Adding to the authors' acknowledgments, I wish to thank my students in the Avian Ecology Laboratory at the University of Missouri—in particular, Judith Toms, she of the degrees in both stats and ecology—whose reviews greatly improved this monograph. Comments from Marissa Ahlering, Andrew Cox, and Ernesto Ruelas Inzunza were helpful to the authors in making this as user-friendly as possible.

John Faaborg

Ornithological Monographs
Volume (2006), No. 59, 1–56

Printed in USA.

MODELING APPROACHES IN AVIAN CONSERVATION AND THE ROLE OF FIELD BIOLOGISTS

Steven R. Beissinger,[1,8] Jeffrey R. Walters,[2] Donald G. Catanzaro,[3] Kimberly G. Smith,[3] John B. Dunning, Jr.,[4] Susan M. Haig,[5] Barry R. Noon,[6] and Bradley M. Stith[7]

[1]*Department of Environmental Science, Policy and Management, University of California, Berkeley, California 94720, USA;*
[2]*Department of Biology, Virginia Polytechnic Institute and State University, Blacksburg, Virginia 24061, USA;*
[3]*Center for Advanced Spatial Technologies and Department of Biological Sciences, University of Arkansas, Fayetteville, Arkansas 72701, USA;*
[4]*Department of Forestry and Natural Resources, Purdue University, West Lafayette, Indiana 47907, USA;*
[5]*U.S. Geological Survey, Forest and Rangeland Ecosystem Science Center, 3200 SW Jefferson Way, Corvallis, Oregon 97331, USA;*
[6]*Department of Fishery and Wildlife Biology, Colorado State University, Fort Collins, Colorado 80523, USA; and*
[7]*Department of Wildlife Ecology and Conservation, University of Florida, Gainesville, Florida 32611, USA*

Abstract.—This review grew out of our realization that models play an increasingly important role in conservation but are rarely used in the research of most avian biologists. Modelers are creating models that are more complex and mechanistic and that can incorporate more of the knowledge acquired by field biologists. Such models require field biologists to provide more specific information, larger sample sizes, and sometimes new kinds of data, such as habitat-specific demography and dispersal information. Field biologists need to support model development by testing key model assumptions and validating models. The best conservation decisions will occur where cooperative interaction enables field biologists, modelers, statisticians, and managers to contribute effectively.

We begin by discussing the general form of ecological models—heuristic or mechanistic, "scientific" or statistical—and then highlight the structure, strengths, weaknesses, and applications of six types of models commonly used in avian conservation: (1) deterministic single-population matrix models, (2) stochastic population viability analysis (PVA) models for single populations, (3) metapopulation models, (4) spatially explicit models, (5) genetic models, and (6) species distribution models. We end by considering the intelligent use of models in decision-making, which requires understanding their unique attributes, determining whether the assumptions that underlie the structure are valid, and testing the ability of the model to predict the future correctly. *Received 30 August 2005, accepted 25 November 2005.*

Resumen.—Esta revisión surgió al reconocer que los modelos juegan un papel cada vez más importante en conservación, pero son raramente usados en las investigaciones realizadas por la mayoría de los biólogos que trabajan con aves. En la actualidad se están creando modelos complejos que involucran mecanismos que podrían incorporar más del conocimiento que han adquirido los biólogos de campo. Estos modelos requieren que los biólogos de campo provean información más específica, utilicen tamaños muestrales mayores y que en algunos casos provean nuevos tipos de datos, como demografía en hábitats específicos e información sobre dispersión. Los biólogos de campo deben apoyar el desarrollo de modelos a través de la prueba de los supuestos claves y la validación de los modelos. Las mejores decisiones en conservación ocurrirán al existir una interacción cooperativa y efectiva entre biólogos de campo, biólogos que realizan modelos, estadísticos y personas que trabajan en manejo.

Comenzamos discutiendo la forma general de los modelos ecológicos—heurísticos o que describen mecanismos, "científicos" o estadísticos—y luego destacamos la estructura, las fortalezas y debilidades y las aplicaciones de seis tipos de modelos que se utilizan

[8]E-mail: beis@nature.berkeley.edu

comúnmente en la conservación de aves: (1) modelos determinísticos matriciales de una única población, (2) modelos de análisis estocásticos de viabilidad poblacional (AVP) para una única población, (3) modelos metapoblacionales, (4) modelos espacialmente explícitos, (5) modelos genéticos y (6) modelos de la distribución de las especies. Terminamos considerando el uso inteligente de modelos en la toma de decisiones, lo que requiere entender los atributos específicos de cada modelo, determinar si los supuestos que subyacen a la estructura son válidos y probar la habilidad del modelo para predecir el futuro correctamente.

Introduction

Avian biologists involved in conservation activities encounter formal mathematical and simulation models ever more frequently and in ever more diverse forms. Models are constructed to act as descriptions of ecological systems (Maynard Smith 1974). As in other sciences, such models have driven the development of certain concepts in conservation biology, such as population viability and metapopulation dynamics. Mathematical and simulation models (hereafter "models") have been used to predict outcomes based on past, current, or projected conditions, and they serve a useful role in synthesizing knowledge and guiding research. To make a model, one is forced to state explicitly the relations between external factors and the state of the system, and this quickly reveals the limits of our understanding. More significantly, models have become important tools that are applied to policy decisions, and their use will continue to expand as desktop computing power grows and user-friendly software makes modeling increasingly accessible. Models, however, are neither a panacea nor the only useful kind of analysis for making conservation decisions. Intelligent use of models in decision-making requires understanding their unique attributes, determining whether the assumptions that underlie the structure are valid, and testing the ability of the model to predict the future correctly.

The present review, and a symposium at an American Ornithologists' Union (AOU) meeting sponsored by the AOU Conservation Committee, grew out of our realization that models play an important role in conservation but are rarely incorporated in the research of most avian biologists. For example, at a recent AOU meeting, only ~4% of 317 papers presented or tested models, compared with ~21% at a meeting of the Ecological Society of America held a few days earlier. Nonetheless, most presenters at both meetings employed a statistical model to test the significance of, or evaluate patterns in, their data. Talking about models with ornithologists evokes strong reactions, as evidenced by the responses of AOU meeting attendees to the question: "What is the first thing that you think of when I say the words 'model or ecological model'?". Answers included "hot air," "money for someone else," "predicting the future," "I go right to the Discussion and hope that they know what they are doing," "people who haven't been in the field enough," "computers," "assumptions and generalizations," "reality?," and "something I don't understand at all."

Models, unlike statistics, are not universally accepted as serious tools by field biologists, perhaps because models are perceived as more difficult to test definitively than other kinds of hypotheses (Aber 1997). When models perform poorly, they are typically modified by changing assumptions or input values, rather than rejected. This reduces confidence in model predictions or conclusions. Models, by their nature, are simplifications of and hypotheses about complex natural systems and can never capture all of a system's dynamics. To avoid oversimplification and to make models more useful in resolving conservation problems, one increases the number of parameters that need to be sampled, but this is likely to result in poor estimation of many parameter values. Poor parameter estimation, like oversimplification, causes field biologists to question the value of models. Because models are simplifications and abstractions of nature, all are expected to be "wrong" to some degree, which makes them obvious targets for attack when they become the basis of controversial decisions—as in the debate over conservation planning for the Northern Spotted Owl (*Strix occidentalis caurina*; Noon and McKelvey 1996a, Noon and Murphy 1997).

Despite these problems, their usefulness for making predictions ensures that models will continue to be important in decision-making. In conservation, predicting the future behavior of a population or system under different management programs is of paramount interest, and no credible prediction is possible without a formal or informal model of the system. Increased use of mathematical modeling in conservation is a manifestation of the maturation of conservation biology as a science. As improved computing power facilitates the development of ever more realistic models, their appeal to field biologists grows, as do demands on field biologists to produce data for parameter estimation. As a result, avian biologists need to be able to understand and evaluate models, and modelers need to interact with field biologists. The opportunity exists for a productive synergism between these two groups to improve conservation decisions.

This monograph reviews a set of models that avian biologists are likely to encounter in conservation. As scientists who are both modelers and field biologists, we offer a perspective on models in conservation that is grounded in theory, application, and natural history. We begin by discussing the general form of ecological models. We then highlight the structure, strengths, weaknesses, and applications of six types of models. Our goal is to present a diverse typology of model applications used in avian conservation and to indicate the kinds of information required for each, so that avian biologists may become more aware of the types of data that will improve the realism, precision, and accuracy of modeling efforts. We conclude by examining how models can be used intelligently to make conservation decisions, and how field biologists and modelers can interact to improve the decision-making process.

The Form of Ecological Models

Modeling natural processes consists of constructing a plausible symbolic representation of the dynamics of a system in the form of mathematical equations or rule sets (Pielou 1981). Ecological models can take many forms, but most models are closely related to *a priori* hypotheses about how a system functions. In this sense, models project the consequences of hypotheses (Nichols 2001). Many types of models are relevant to avian research, including those that address optimal foraging (Stephens and Krebs 1986), life history (Charnov 1993), population dynamics (Maynard Smith 1974, Caswell 2001), and game theory (Maynard Smith 1982). Models come in many shapes and sizes, however, and there are multiple typologies of ecological models relevant to avian ecology (Jorgensen and Bendoricchio 2001, Nichols 2001, Gertseva and Gertseva 2004). One typology classifies models as either heuristic or mechanistic (Pielou 1981). Heuristic models capture the essence of a system using a few, sometimes abstract, variables to predict future system behavior, but such models provide no causal explanation for the model outcomes expected to occur. In contrast, mechanistic models are designed to capture the key processes and relationships among variables as they exist in nature, and to provide an explanation for expected outcomes. Models used to project forest growth illustrate this contrast. Traditional models of forest growth and yield are heuristic—they use regression equations to predict future forest growth on the basis of past, empirically estimated, relationships (Pielou 1981, Davis et al. 2001). The form of these equations and the regression coefficients may not have direct biological interpretations. In contrast, forest succession models are mechanistic and attempt to project future tree structure and composition on the basis of known aspects of the survival process, physiological tolerance to shading, competitive ability to obtain light and water, and recruitment dynamics based on dispersal and propagule number (Shugart 1984, Huston 1992). These models require an understanding of mechanistic processes, and therefore depend on experimental studies of processes by field biologists. All the model types we discuss in this monograph contain both heuristic and mechanistic elements, but they vary in relative emphasis. As a model's complexity increases, it can incorporate more key processes and becomes more mechanistic. The population models we discuss generally represent an increasingly complex, increasingly mechanistic series (deterministic single-population, stochastic single-population, metapopulation, spatially explicit).

Another useful characterization of models distinguishes between scientific and statistical models (Nichols 2001, Williams et al. 2002). Scientific models are used to project the consequences of a hypothesis of how a system works, usually expressed in mechanistic terms. For example, a scientific model may predict how the reproductive output of a certain bird species will be affected by weather, habitat, and their interaction during the breeding season (Franklin et al. 2000). Statistical models (and hypotheses) follow logically from their corresponding scientific models; they project how measurable quantities or data should appear if consistent with the scientific model.

A final useful partition of ecological models is based on whether their objective is to describe biological processes or to facilitate management decisions (Taylor 1995). Biological models are often mechanistic, but are not constrained in their design (i.e. their input parameters and outputs) to reflect the concerns of managers. For example, a key parameter whose value would trigger a management response is not required in optimal-foraging models. Management models, in contrast, are specifically designed to facilitate decision-making even in the context of great uncertainty (Williams et al. 2002, Dale 2003). Such models should be based on parameters that are easily estimated and relevant to managers. Differences in the construction of biological and management models may limit the value of some biological models for management decisions. All the models we examine below exhibit characteristics of both biological and management models.

It is important to recognize that the model types discussed above are not mutually exclusive and that we have not exhaustively discussed all possible characterizations. In addition, most models treated in this monograph can be classified within one or more typologies. For a discussion of other useful characterizations of model types relevant to management and conservation, see Williams et al. (2002).

Inherent differences between the predictions or outputs of a biological model and the factors affected by management often inhibit conservation decisions. In the policy arena, our limited understanding of ecological processes and the uncertainties associated with projected consequences of environmental change are often used to inhibit action and protect the status quo. Those who oppose costly conservation decisions find complex models based on simplifying biological assumptions to be easy targets. What they fail to recognize, however, is that any forecast of the consequences of an action is inherently based on some model of how the system works. This includes the prediction of "no adverse consequences," which also is based on some mental construct of the system. The great advantage of an intelligible mathematical or simulation model is that it makes explicit the structure, assumptions, and relationships among the system's variables. For example, a population model can include various factors believed to affect survival and fecundity, such as habitat quality, habitat distribution, annual variation in food supply, and predation rate. With an explicit model, one can assess the importance of different assumptions about relationships among factors, incorporate new information about processes, and test and refine the ability to predict population dynamics. Models may, unfortunately, also lead to wrong decisions when they do not incorporate important processes or unintentionally use incorrect parameter estimates (Emlen 1989). The critical question is to what extent a conservation decision will benefit from considering the results of a particular model. The answer depends on the model's attributes.

In the following six sections, we discuss the attributes of biological models that have been transformed into management models and describe their structures, assumptions, and uses in conservation. We conclude with a return to the general topics of (1) intelligent use of models in conservation and (2) interactions of field biologists with modelers.

Deterministic Single-population Matrix Models

Deterministic single-population matrix models (hereafter "matrix population models") are among the simplest demographic models. They are descriptions of population dynamics consisting of a set of equations (one for each age or stage class) that predict population size at time $t + 1$ from information on the survival, growth, and reproduction of individuals at time t. The equations are often formulated into a matrix, which is a rectangular array of numbers or symbols. Matrices have mathematical properties discussed below that are directly related to important measures of population dynamics, and they are very convenient for analysis with computers. Matrix population models have long been used in ecology (Leslie 1945, 1948; Lefkovitch 1965; Getz and Haight 1989; Caswell 2001), and they have been employed to make management recommendations for birds since the late 1960s. Applications have included endangered species, such as the Whooping Crane (*Grus americana*; Miller and Botkin 1974) and California Condor (*Gymnogyps californianus*; Mertz 1971), as well as seabirds (Leslie 1966) and game birds (Geis et al. 1969, Anderson 1975, Nichols et al. 1995). They became further established as a tool for making population management decisions with the incorporation of sensitivity analysis (e.g. elasticity, discussed below) (Crouse et al. 1987, Crowder et al. 1994, Doak et al. 1994, Silvertown et al. 1996).

It is not appropriate to use matrix models to make long-term population projections unless it is assumed that long-term averages for the vital rates will experience little change. Matrix models can be used to estimate the geometric rate of population growth (lambda, or λ); to indicate population characteristics, such as the distribution of individuals among age classes (stable age distribution) or the reproductive value of age classes; and (as it is often used in conservation) to evaluate the influence of demographic rates on population change (sensitivity analysis) (McDonald and Caswell 1993, Caswell 2000). If the relationship between proposed management actions and demographic rates can be determined, the ability of different management actions to produce positive population change under current conditions can be compared. Thus, a matrix population model might indicate whether management actions that increase nesting success will be more effective than management actions that increase adult survival, even though the model may not accurately project population size over time.

Methods other than matrix population models can be used to estimate lambda. The traditional approach uses the geometric mean of the ratios of population size (N) estimates for consecutive periods ($\lambda = N_{t+1}/N_t$) calculated from a time series (Morris and Doak 2002, Williams et al. 2002), whereas a newer approach is based on mark–recapture data using the temporal symmetry method of Pradel (1996). The temporal symmetry method fits mark–recapture models for open populations to capture data viewed simultaneously in a forward and backwards manner (Nichols and Hines 2002, Williams et al. 2002). When the capture data are viewed in a forward manner, the model estimates annual local survival (ϕ_t; the probability that an individual in the population in year t will survive and remain in the population in year $t + 1$); when viewed backwards, the model estimates the seniority probability (γ_t; the probability that an individual did not enter the population between years t and $t - 1$). Lambda at time t (λ_t) is estimated by ϕ_t / γ_{t+1}. Neither of these approaches is reviewed here, because they are essentially statistical rather than demographic models and as yet do not allow incorporation of mechanistic processes (Nichols et al. 2000, Nichols and Hines 2002, Williams et al. 2002). Nevertheless, they can have important conservation applications (Dreitz et al. 2002, Cam et al. 2003, Franklin et al. 2004). Both approaches calculate direct estimates of lambda, whereas matrix population models yield an asymptotic rate of population change based on an expected rate of population change modeled from an observed set of vital rates (Caswell 2001). Sandercock and Beissinger (2002) showed that all three approaches can yield consistent estimates of lambda. Matrix population models have the advantage of incorporating more biological processes, but the disadvantage of sometimes yielding unrealistic values of lambda, because of either poor estimates of vital rates or a mismatch between the actual and asymptotic stable age distribution.

Structure of Matrix Population Models

Matrix population models are the least data-demanding of the models reviewed here (Fig. 1), though years of field study are needed to construct and parameterize them well. These models require only: (1) an understanding of age, stage, or social structure (to determine ages or stages for analysis); (2) identification of age or stage of first reproduction; and (3) estimates of reproductive success and survivorship for the different ages or stages. Ideally the decision of how many stages to use and their composition should be closely tied to variation in demographic parameters (Sauer and Slade 1987, Caswell 2001, Pfister and Wang 2005). In practice, however, ages or stages of long-lived organisms are often collapsed into a few classes because field studies are rarely long enough to measure age-specific rates. Nevertheless, many

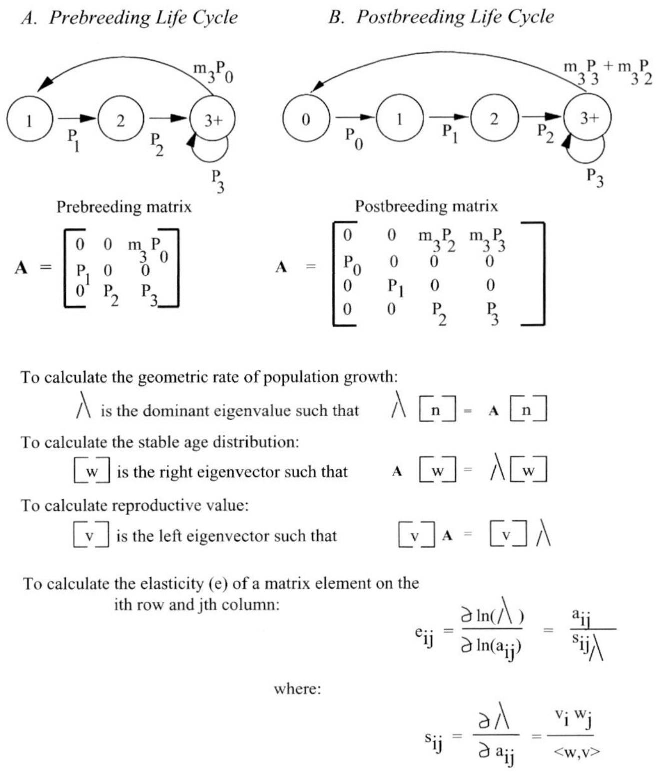

FIG. 1. Matrix population models for (A) prebreeding and (B) postbreeding life cycles, and elasticity calculations (modified from Beissinger and Westphal 1998). Life-cycle diagrams are presented with nodes for four age classes: 0 (juveniles that are young of the year), 1 (subadults, hatched in the previous year, and up to 1 year of age), 2 (subadults from 1 to 2 years of age), and 3 (adults ≥2 years old). Individuals of each age class (indicated by subscripts) survive with a probability of P, but only adults reproduce with a fecundity of m. No juvenile node occurs in the prebreeding life-cycle diagram, because censuses are conducted just prior to the breeding season, when surviving young have already become 1-year-olds at a juvenile survival probability of P_0; this appears in the prebreeding life-cycle diagram in the first row of the matrix. The postbreeding life cycle and matrix have an extra node and row, respectively, for juveniles, because censuses are conducted immediately after the breeding season. The first row of the postbreeding matrix models reproduction as a product of fecundity and subadult (P_2) or adult survival (P_3). The reason is that only a portion of these individuals will survive from the postbreeding census until the beginning of the next breeding season, increase in age by one year, and then reproduce. Basic matrix calculations can easily be solved using commercially available mathematics software by one of two methods. The power method (McDonald and Caswell 1993) raises the matrix to a high power (e.g. $\mathbf{A}^{128}$), which causes the rows and columns of the matrix to converge on proportions that do not change. The stable age distribution is calculated from the proportions of the coefficients of the columns; the coefficients of the rows converge on reproductive value, which is found by dividing each coefficient by the value of the first coefficient in its row. Lambda (λ), the geometric rate of annual population growth, is found by dividing any cell in $\mathbf{A}^{128}$ by the same cell in $\mathbf{A}^{127}$. Alternatively, matrix algebra (Caswell 2001) can be used to calculate the stable age distribution (right eigenvector), reproductive value (left eigenvector), and lambda (dominant or largest positive eigenvalue or latent root of the matrix). Elasticity requires calculating the derivative of the log of lambda with respect to the log of a matrix element (a_{ij}) situated in the ith row and jth column of the matrix. The formula first requires calculation of the sensitivity (s_{ij}) of the element, which is the partial derivative of lambda with respect to the element. The term ($v_i w_j$) is the product of the reproductive value and stable age distribution associated with the row and column address of a particular element, whereas the term $\langle w,v \rangle$ is the sum of all the element by element products of the stable age distribution and reproductive value.

birds have complex social structures, which may include subadults that have yet to reach the age of first breeding, nonbreeding adults old enough to breed (e.g. helpers), senescing adults, etc. An understanding of social structure is critical to determining the underlying structure of the matrix (McDonald and Caswell 1993, Harcourt 1995). Estimates of age- or stage-specific rates of fecundity and survivorship are usually the most important components of matrix population models, and the accuracy of their estimation is likely to have a large effect on model outcomes. Most matrix population models of vertebrates are constructed only for females, because male fecundity is often unknown. Thus, rates are usually expressed on a per-female basis. Nevertheless, two-sex models can be built (Caswell 2001).

Caswell (2001) and McDonald and Caswell (1993) present detailed treatments of the structure of matrix population models and the calculations involved, so we summarize these only briefly here. The underlying structure of a matrix model is the life cycle of the organism, which can be depicted in a diagram as a set of nodes for stages and arcs for transitions between stages (Fig. 1). In the matrix, a row is required for each stage- or age-class of the modeled population, and columns track the contribution of each stage or age to the row. The first row of the matrix accounts for rates of production and recruitment of young into the population, and the other rows present rates of survival within stages or the transition from one stage to another. Correct construction of matrix models critically depends on the time at which the population is censused in relation to the time of reproduction (Noon and Sauer 1992).

Many matrix analyses are easily done using commercially available mathematics software (e.g. MATLAB, MATHEMATICA, and MATHCAD) requiring only the input of the matrix followed by a few commands (Caswell 2001, Morris and Doak 2002). Once the matrix is constructed, it is easy to calculate (1) the annual rate of population growth (lambda), which describes the rate of change for each stage-class and the population once the population has reached a stable age distribution; (2) the stable age distribution, which is the proportion of individuals in each age- or stage-class if the rates of survival and reproduction remain unchanged; and (3) reproductive value, which is the expected contribution of an individual in a particular age- or stage-class to future population growth. Nevertheless, the mathematical terminology and processes involved in matrix algebra can be confusing to those unfamiliar with them (Caswell 2001). An introduction to the calculations is provided in Figure 1; McDonald and Caswell (1993) and Morris and Doak (2002) give more detailed, but readable, presentations.

Sensitivity can be analyzed in several ways (Caswell 2000, 2001), but elasticity is employed most frequently (Fig. 1). Elasticity is the proportional change in annual population growth (lambda) resulting from a proportional change in a matrix element (de Kroon et al. 1986; de Kroon et al. 2000). Elasticity values for the elements of matrix population models have the convenient property of summing to 1.0 and give a proportional contribution to the total sensitivity of lambda. Thus, small changes in the vital rates that compose a matrix element having a large elasticity value will produce a much greater effect on the rate of population growth than equally small changes in the vital rates that comprise a matrix element having a small elasticity value. However, the first row of matrix elements is calculated as a product of both survival and fecundity (Fig. 1). Thus, lower-level elasticities must be calculated to partition elasticity among survival and fecundity (Wisdom and Mills 1997, Caswell 2001). When a vital rate is found on more than one matrix element, such as adult survival, lower-level elasticity values can be added to obtain a total elasticity value for the vital rate, but the elasticity values will no longer sum to 1.0. For all demographic subcomponents found only on the same element, lower-level elasticity values will take on the same value. In birds, for example, fecundity is frequently calculated as a product of the proportion of females breeding, proportion of nests fledging young, and number of female young fledged per successful nest. This can limit the usefulness of elasticity in instances where the goal is to understand the effect of changing a particular subcomponent of reproduction (e.g. proportion of nests fledging young).

Two less frequently used but equally valuable approaches to analyzing the sensitivity of matrix population models are the life-table response experiment (LTRE) and life-stage simulation analysis (LSA). The LTRE is an analytical approach that extends elasticity by incorporating specific changes to vital rates (Caswell 1996b, 2000). It compares the contributions of specific changes in particular vital rates with differences in lambda between a "mean" or "control" matrix and a "perturbed" or "treatment" matrix. The LTRE does this by taking the product of the change in the vital rate caused by perturbation and the sensitivity of lambda to changes in that rate (Caswell 2000, Mills and Lindberg 2002). Thus, LTRE can compare the effects on lambda of the overall contribution of changes in vital rates. Although frequently used to analyze plant population dynamics (Bruna and Oli 2005, Brys et al. 2005, Griffith and Forseth 2005), LTRE has rarely been applied to birds; for a notable exception, see Cooch et al. (2001). Life-stage simulation analysis is a simulation-based approach to sensitivity analysis. It evaluates the effect on lambda of changes in vital rates by constructing hundreds of replicate matrices randomly drawn from specified distributions for each vital rate (Wisdom and Mills 1997, Wisdom et al. 2000). Lambda is calculated for each matrix, and the coefficient of determination (r^2) between the value of each vital rate and lambda is found; this reveals which vital rate accounts for the greatest variation in the growth rate for all simulations when rates change simultaneously.

Use of Matrix Population Models in Conservation

Use of deterministic matrix population models in management decisions has grown rapidly since the development of elasticity analysis. Matrix analyses were used to estimate rates of population growth for loggerhead sea turtles (*Caretta caretta*) to compare the effects of different management options; these indicated that turtle-excluder devices are more effective than *in situ* and *ex situ* egg protection (Crouse et al. 1987, Crowder et al. 1994, Grand and Beissinger 1997). Similarly, analyses of elasticity were used to evaluate the effects of proposed management techniques on lambda in Red-cockaded Woodpeckers (*Picoides borealis;* Heppell et al. 1994) and Greater Prairie Chickens (*Tympanuchus cupido pinnatus;* Wisdom and Mills 1997). Matrix models were applied to Marbled Murrelets (*Brachyramphus marmoratus*) to estimate lambda and to compare the ratio of juveniles to after-hatch-year birds, which is an indirect estimate of recruitment (Peery et al. 2006a), with model estimates of productivity needed for stable populations (Beissinger 1995a, Beissinger and Nur 1997, Peery et al. 2006b). Matrix models were used to evaluate acceptable rates of mortality required to re-establish viable populations of California Condors (Meretsky et al. 2000), to analyze risks of decline in and effects of long-line fisheries on Amsterdam Albatrosses (*Diomedea amsterdamensis;* Inchausti and Weimerskirch 2001), to specify the effects of feral cats on island-nesting Black-vented Shearwaters (*Puffinus opisthomelas;* Keitt et al. 2002), and to determine habitat-specific differences in population growth in Peregrine Falcons (*Falco peregrinus;* Kauffman et al. 2003) and ptarmigans (*Lagopus* spp.; Sandercock et al. 2005). Matrix models have also been used to compare the efficacy of population control options for Snow Geese (*Chen caerulescens;* Mills and Lindberg 2002) and Brown-headed Cowbirds (*Molothrus ater;* Citta and Mills 1999). In these two applications, as well as in the Greater Prairie Chicken example (Wisdom and Mills 1997), LSA was employed as an alternative form of sensitivity analysis to indicate how potential variation in vital rates was affecting the population dynamics. Hoekman et al. (2002) also used multiple approaches to explore the sensitivity of Mallard (*Anas platyrhynchos*) life-cycle stages. Cooch et al. (2001) used LTRE to analyze the demographic responses of Lesser Snow Geese (*Anser c. caerulescens*) to changes in food abundance.

Although matrix population models require the least data of all the modeling approaches reviewed here, developing accurate and precise estimates for vital rates and constructing the matrix is not a trivial matter. First, accurate estimates of fecundity and survival based on long-term studies are rarely available for endangered species (Noon and Sauer 1992, Beissinger and Westphal 1998). Survivorship is often the most difficult demographic rate to measure accurately, because it must be distinguished from the probability of resighting (Nichols 1992). Complex statistical techniques have been developed to yield accurate estimators of survival and to test for differences among age- or stage-classes (Lebreton et al. 1992). Unless the probability of resighting is very high, they usually require large samples of marked individuals followed over at least three years to estimate the probability of resighting and survivorship for a single year. It is unfortunate that survival estimates can be problematic, because sensitivity analyses suggest that population change in most long-lived vertebrates is often most affected by changes in adult survivorship (Lande 1988b, Pfister 1998). For example, Sæther and Bakke (2000) compared elasticities of 49 species of birds and found that adult survival tended to have the greatest effect on population growth, though the contribution of fecundity increased with increasing clutch size. Second, the structure and values in the matrix depend on whether it is parameterized as a prebreeding or postbreeding model, or as a birth-flow (births occur continuously over the projection interval) or birth-pulse (births occur during a short breeding season within the interval) model (Noon and Sauer 1992, Caswell 2001). The largest differences are in the calculation of realized fecundity, in the first row of the matrix (Fig. 1).

Matrix population models are often employed to recommend management strategies, but results need to be interpreted sensibly. These models are simple descriptions of population dynamics. Matrix population models rarely attempt to incorporate processes that produce the observed survival and reproductive rates, and they often ignore the effects of uncertainties in parameter estimation (Wisdom et al. 2000), covariation among vital rates (van Tienderen 2000), density dependence (Grant and Benton 2000), and stochasticity (de Kroon et al. 2000). This limits the appropriate uses of these models in several ways. First, sensitivity analyses can indicate which vital rates or stages most affect model outcomes or require more study for better parameter estimation, and they can be used to identify the management strategy that may lead to the fastest population recovery. However, sensitivity analyses do not indicate which factors are causing populations to decline (Green and Hirons 1991, Beissinger and Westphal 1998). The latter information requires comparative, behavioral and experimental approaches (Lande 1988b, Caughley and Gunn 1996, Norris 2004, Peery et al. 2004), or comparison of model trajectories with real population trajectories (Hitchcock and Gratto-Trevor 1997, Peery et al. 2004). In addition, recovery will require more time in populations that are declining because of limiting factors that operate on elements with low sensitivity. Moreover, sensitivity analysis does not indicate the degree to which it is feasible to influence a vital rate with different management options and does not incorporate the cost per unit change in lambda of particular management options, a metric that may be important to decision-makers (Nichols and Hines 2002).

Elasticity analysis has also been criticized because it can lead to inconsistencies in evaluating demographic rates, depending on how they are scaled (Link and Doherty 2002). The proportional changes of an elasticity analysis are unitless and calculated on a log–log scale. However, demographic rates are measured on different scales (e.g. annual survival varies from 0 to 1, whereas annual fecundity values may exceed 10 in ducks and thousands in fishes and invertebrates). Link and Doherty (2002) argue that the log–log scale is mainly appropriate for transformation of probabilities (bounded by 0 and 1), but should be replaced with variance-stabilized transformations when analyzing matrices that contain elements outside that range. See Doherty et al. (2004) for a recent application of this approach with a seabird. Because of these and other limitations, Norris and McCulloch (2003) and Norris (2004) recommend a greatly reduced role for elasticity analysis in influencing management decisions, and instead suggest analyzing the feasible effects of management on vital rates and the resulting population growth rates. This kind of scenario analysis provides a direct way to evaluate the effects of management on population growth. However, the validity of the insights depends upon the ability of management to achieve the demographic improvements specified in the scenarios being evaluated.

A second limitation on use of sensitivity analyses is that they are not value-free, but depend on the vital rates used in the matrix (Caswell 1996a). Changes in values used in the matrix may sometimes shift the rankings of elasticity results (Mills et al. 1999, Wisdom et al. 2000). Furthermore, if a demographic rate is depressed by the effects of a limiting factor, its matrix element(s) will have a smaller elasticity value. This does not mean that factors or stages associated with this element are less important for management than other elements. For example, elasticity values for the Mariana Crow (*Corvus kubaryi*) on Guam indicated an overwhelming importance of adult survivorship, but successful reproduction has rarely occurred during the past decade because of nest predation by the brown tree snake (*Boiga irregularis*) (National Research Council 1996, Beissinger 2000). The model was based on low fecundity values observed under current conditions, not the presumably higher but unobserved values that prevailed before the invasion of the snakes. Recovery strategies that addressed only adult survivorship would miss the main cause of decline, which was poor reproduction caused by snake predation on eggs and nestlings.

A third limitation is that population estimators derived from deterministic matrix analyses assume that demographic rates are constant, an assumption that is violated to some extent in all matrix applications. For example, ecosystems can experience severe environmental fluctuations on relatively short cycles as compared with the generation time of birds (e.g. El Niño), and these can greatly affect demography (Beissinger 1986, 1995b; Grant 1986). The effects of violating this assumption depend on how much variation in demographic rates occurs from year to year (Wisdom and Mills 1997, Caswell 2001). In addition, positive population growth rates do not indicate that a population is secure, because environmental variation and catastrophes can cause large fluctuations in population size and result in a high chance of extinction (Goodman 1987, Tuljapulkar 1989, Lande 1993, Mangel and Tier 1994, Sæther et al. 2005). Stochastic single-population models are a more complex alternative to deterministic models, and they are more realistic in that they incorporate variation in demographic rates.

Stochastic Population Viability Analysis Models for Single Populations

Stochastic demographic models of single populations are another way to describe population dynamics. These models usually do not incorporate the processes that produce particular rates of survival and reproduction, but to some extent they simulate the effects of these processes by including annual variation in vital rates or lambda. Stochastic models are used to make long-term population projections, especially in the context of population viability analysis (PVA; Shaffer 1981, 1987; Beissinger and McCullough 2002). A population is projected for 50, 100, or more years into the future by allowing lambda or demographic rates to change for each time step (usually one year); the method is Monte Carlo, which samples rates randomly from predetermined distributions (Fig. 2). Each run of a stochastic model follows a unique trajectory and yields a different ending population size, because the lambdas or demographic rates change randomly with each time step. Whereas a deterministic matrix population model produces a single-population projection that changes at the rate of lambda, stochastic single-population models yield probabilistic results. Models must be run 500–1,000 times to explore the full range of parameter values so that results portray the distribution of possible ending population sizes (Harris et al. 1987, Burgman et al. 1993, Morris and Doak 2002).

Results from stochastic single-population models can be summarized in several ways (Burgman et al. 1993, Beissinger and Westphal 1998). The most common for evaluating population viability is the proportion of runs that end at population size zero ("extinction" rate), or at a small size, such as ≤25 individuals ("quasi-extinction" rate). No standard interval or extinction rate defines a viable population, but intervals of 50–100 years and extinction rates <5% are commonly used (Ralls et al. 2002, Reed et al. 2002). A second result computes the expected "time to extinction," which is the mean or median year of extinction for populations that went extinct. A third approach is to calculate the mean and standard deviation for population size at each interval. Perhaps the most complete descriptor of model results is to plot the cumulative probability function for ending population size (Fig. 2D). This is known as the quasi-extinction function and is a basic form of risk analysis (Ginzburg et al. 1982, Burgman et al. 1993). The minimum viable population size (MVP) can be found by changing the initial population size to find the smallest size that had a 95% chance of remaining extant at the end of the period evaluated in the simulation (Shaffer 1981, Soulé 1987). Minimum viable population sizes from models are rarely used in management contexts, because of the difficulty in accurately forecasting risks of extinction (Ralls and Taylor 1997, Reed et al. 2002), though estimates sporadically appear in the published literature (Harcourt 2002, Reed et al. 2003).

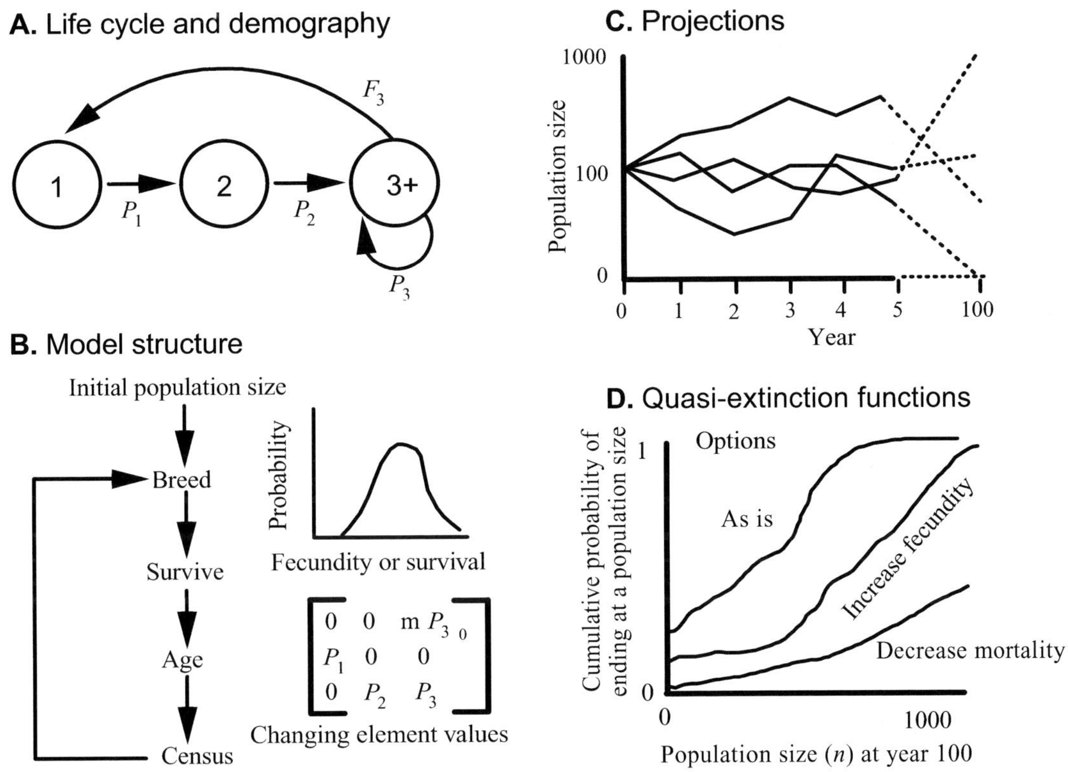

FIG. 2. Simplified example of the structure and outcomes from a stochastic single-population PVA model (from Beissinger and Westphal 1998). (A) The prebreeding life-cycle diagram gives the model age or stage structure (see Fig. 1). (B) The basic flow of events of the prebreeding life cycle structures the model. Rates for fecundity (m) and survival (P) are randomly chosen anew for each time step (from predetermined distributions) and used in matrix calculations to project the population size at the next census (i.e. year). (C) The initial population is projected over many years in a single iteration of the model, as shown by any of the trajectories. Projections are repeated 500–1,000× to simulate different possible population trajectories. (D) Outcomes for various management options (e.g. current conditions or "as is," increasing fecundity, or decreasing mortality) are determined, such as the quasi-extinction function, extinction rate, time to extinction, and average population size. The quasiextinction function incorporates the population size from all projections at a specified interval and is determined by calculating the cumulative probability for populations ending less than or equal to a particular size at the specified time interval. The extinction rate is where the function intersects the y-axis.

STRUCTURE OF STOCHASTIC POPULATION VIABILITY ANALYSIS MODELS FOR SINGLE POPULATIONS

Stochastic single-population PVAs can be built as either unstructured or structured models. Unstructured models use estimates of the mean and variance of lambda to project population fluctuations into the future by employing Monte Carlo techniques to sample lambdas randomly from predetermined distributions. They contain no structure or information about individuals (e.g. ages, stages, or sizes). Unstructured models are usually parameterized from a time series of counts or estimates of population size for consecutive periods ($\lambda = N_{t+1}/N_t$) or from the log of the counts (Morris and Doak 2002, Williams et al. 2002); the latter measure is used to calculate the stochastic rate of population growth (μ), which is equivalent to r (exponential rate of increase) rather than lambda (geometric rate of increase). The mean and variance of μ can be used in analytical models to yield estimates of the probability that a population of a given size will reach any size threshold at some number of years in the future based on the diffusion approximation (Dennis et al. 1991, Foley 1994, Holmes 2004). Although the concepts behind these applications of stochastic calculus are sophisticated, Morris and Doak (2002) provide very accessible details for doing the calculations.

Parameterizing structured single-population models requires about twice as many rates to be estimated than are needed for matrix population models (Beissinger and Westphal 1998), though many of the additional rates can be derived from the same field data. The basic skeleton of structured stochastic population models is similar to that of deterministic matrix models, so they require the same estimates of mean age- or stage-specific survival and fecundity. In addition, to model the effects of demographic and environmental stochasticity, models require estimates of variance in fecundity and survival for each age- or stage-class. Demographic stochasticity refers to chance events in birth or death rates attributable to population size. For example, the probability of survival for a particular age-class may be 60%, but each individual either lives or dies. The variability that results from this process represents demographic stochasticity. Demographic stochasticity can be modeled by choosing the number of draws from a binomial distribution equal to the population size, with probabilities chosen from a range of realistic values for the demographic rate of interest. Environmental stochasticity refers to changes in demographic parameter values caused by fluctuations in the environment that affect all individuals similarly (e.g. annual variation in weather or resources; Lande 2002). Accurate modeling of environmental stochasticity is difficult; it requires knowledge of the relationship between environmental conditions and vital rates and their variance over the full range of conditions encountered.

Stochastic single-population models should also include additional complexities, namely some method of incorporating density dependence and carrying capacity and the frequency and effects of catastrophes. Carrying capacity sets an upper limit on how large populations can grow and density dependence affects population growth rates. Models without such limits often overestimate population persistence and underestimate maximum growth rates (Ginzburg et al. 1990, Sabo et al. 2004). Various density-dependent functions can be used to model the effects of approaching carrying capacity, or a population ceiling size may be designated that acts as an upper boundary for population trajectories (Burgman et al. 1993, Foley 1994, Sabo et al. 2004). Structured stochastic single-population models can also incorporate stochastic processes that may occur when populations become small, such as Allee effects (Courchamp et al. 1999, Stephens et al. 1999) or inbreeding (Charlesworth and Charlesworth 1987, Lande 2002). Catastrophes are a form of environmental variation that can be distinguished from environmental stochasticity by the magnitude of effects on demography. They may result in large population declines and can greatly increase the chance of extinction (Mangel and Tier 1994), but are not necessarily rare events. Physical forces that strongly affect demography, such as droughts and floods, can occur in short intervals (e.g. 5–7 years) and in predictable cycles (Beissinger 1986). Determining the predictability of important environmental factors (Colwell 1974, Stearns 1981, Beissinger and Gibbs 1993) is required to assess whether models should allow demography to vary entirely stochastically or with an underlying form of quasiperiodicity.

Conservation Use of Stochastic Population Viability Analysis Models for Single Populations

Stochastic single-population models have frequently been used to estimate the likelihood of extinction for wild populations and, on that basis, to make management recommendations. Shaffer (1981, 1983) first used PVA on Grizzly Bears (*Ursus arctos*) in Yellowstone National Park, though stochastic population models had occasionally been used to explore the demography of endangered and harvested birds (Nichols et al. 1980, 1995). A structured stochastic single-population model was used to predict that Red-cockaded Woodpeckers in the Georgia Piedmont and Piping Plovers (*Charadrius melodus*) in the Great Plains had high likelihoods of extinction within 100 years (Ryan et al. 1993, Maguire et al. 1995). These models have also been used to evaluate the necessity of removing Puerto Rican Parrots (*Amazona vittata*) from the wild for captive breeding (Lacy et al. 1989), the viability of captive and reintroduced populations of Bearded Vultures (*Gypaetus barbatus*) in the Alps (Bustamante 1996), threats to an island population of Capricorn Silvereyes (*Zosterops lateralis chlorocephala*; Brook and Kikkawa 1998), the effect of fisheries on Amsterdam Albatrosses (Inchausti and Weimerskirch 2001), and risks to Balearic Shearwaters (*Puffinus mauretanicus*; Oro et al. 2004). Unstructured single-population models have been used to examine life-history correlates of time to extinction in birds (Sæther et al. 2005); and a variation of this approach, the population prediction interval (Sæther and Engen 2002), has been used to examine risks to a declining population of Barn Swallows (*Hirundo rustica*) in Europe (Engen et al. 2001).

Like deterministic matrix models, structured stochastic models of single populations are most useful for examining the effects of different management options (Fig. 2D). This is often done by comparing population projections of different management scenarios and by sensitivity analysis, which, unlike deterministic matrix models, cannot be calculated analytically. The conventional method of sensitivity analysis for stochastic PVA models is to determine the change in the probability of extinction (or another model output) in relation to fixed-percentage changes in a model parameter (Beissinger 1995b). This involves adjusting model parameters one at a time, conducting many iterations for each new parameter set, and comparing the results with the average outcome, which is calculated by setting all input parameters to their mean value. For example, Reed et al. (1998) constructed a structured stochastic PVA model for the Hawaiian Stilt (*Himantopus mexicanus knudseni*) and found that populations were unlikely to go extinct. Sensitivity analysis indicated that model outputs were most affected by changes in nesting failure and adult survival—factors that managers could minimize by maintaining predator control and limiting water-level fluctuations. Furthermore, the results were insensitive to changes in the probability of catastrophe and density dependence. In addition to conventional sensitivity analysis, sensitivity can be evaluated by examining the effect on model outputs of changing each model parameter by a fixed percentage of its range (relative sensitivity), and by analyzing the impacts of input parameter values on the likelihood of extinction or quasi-extinction using logistic regression (Cross and Beissinger 2001). The latter approach is perhaps the most useful, because it can examine interactions among input parameters and changes to model structure. Cross and Beissinger (2001) recommended employing several kinds of sensitivity analyses to understand how input parameters affect model outcomes, but suggested comparing management scenarios to evaluate the efficacy of potential management options.

Stochastic models provide a more flexible approach than deterministic ones, because demographic rates need not be restricted to current conditions. For example, the effects of silvicultural practices on a threatened mammal were examined to estimate how the likelihood of extinction was affected by changes in carrying capacity (Lindenmayer and Possingham 1996). Similarly, stochastic modeling examined how the interval between low water conditions affected the viability of Snail Kites (*Rostrhamus sociabilis*) in the Florida Everglades (Beissinger 1995b). Demographic rates were partitioned among different environmental states (i.e. drought, lag, and flood years), and the periodic sequence of environmental states dictated vital rates.

Results from stochastic single-population models should be interpreted carefully, of course. Although PVA models supplied with adequate data have the potential to track the average short-term trajectory of a population with accuracy (Brook et al. 2000b), these models are unlikely to produce accurate predictions of the likelihood of extinction (Ludwig 1996, 1999; Beissinger and Westphal 1998; Groom and Pascual 1998; Fieberg and Ellner 2000). First, demographic data are often inadequate, and estimates of vital rates are frequently imprecise and are based on studies too limited in duration to estimate properly the mean and variance. Analyses of long-term data sets of a variety of animals have shown that asymptotic estimates of the variance in population size, if they occur at all, may require at least 8–20 years of sampling (Pimm and Redfearn 1988, Pimm 1991, Ariño and Pimm 1995) Reasonable estimates of variance in vital rates for birds probably require at least several generations of study, which could easily exceed 10–20 years for many species. Second, estimates of variance in vital rates and lambda derived from field studies include sampling error, which should be discarded because it is caused by errors in parameter estimation, as well as the annual (environmental) variation of interest (Nichols et al. 1994, Staples et al. 2004). Studies of Semipalmated Sandpipers (*Calidris pusilla*) found that sampling error contributed nearly as much to the total variance in annual survivorship as temporal variation (Hitchcock and Gratto-Trevor 1997). Methods are available to partition a vital rate's total variation into components caused by sampling error and process error (Gould and Nichols 1998, White et al. 2002).

Third, the primary prediction from stochastic single-population models, the probability of extinction, is difficult to validate because these models incorporate stochastic processes. We cannot know which of the hundreds of simulated population trajectories will most resemble the trajectory of the actual, unreplicated population. Comparing the model's average population projection with a time series of historical population trends provides a way to examine how well the model captures the short-term dynamics of the system (Brook et al. 2000b). But it does not verify the value of stochasticity or other variables used in the model that are responsible for differences among model runs. Furthermore, acceptable levels of risk always occur at one tail of the distribution of possible outcomes (e.g. <5% chance of extinction in 100 years), which makes the accuracy of predictions particularly difficult to evaluate (Ludwig 1996, 1999). Model assumptions or secondary predictions, such as population size or estimates for means and variances of vital rates, are more feasible to test (McCarthy et al. 2001, Lindenmayer et al. 2003). Fourth, differences in model structure can have strong effects on management recommendations resulting from model output. Models with different structures can reproduce the same population dynamics in the absence of management, but may predict different effects of management regimes (Pascual et al. 1997, Gerber and VanBlaricom 2001, La Montagne et al. 2002). Also, different computer programs can produce different estimates of viability from the same data because of differences in the ways of modeling density dependence and sex ratio (Mills et al. 1996, Brook et al. 2000a). Incorporating density dependence may increase model realism, but requires additional validation to determine whether density dependence is modeled accurately. Even when data appear to be adequate, stochastic single-population models can make large errors in estimating the rate of extinction (Taylor 1995, Ludwig 1999, Belovsky et al. 2000, Fieberg and Ellner 2000).

Taken together, these concerns strongly suggest that one should place limited confidence in the extinction estimates generated by these models. Stochastic single-population models are best employed in conservation decisions by comparing the differences in relative rates among management options incorporated into the model rather than basing policies on the absolute rates of extinction predicted (Beissinger and Westphal 1998, Reed et al. 2002, McCarthy et al. 2003, Lotts et al. 2004). This was the basis for comparing conservation strategies based on different sizes and configurations of reserves for the Northern Spotted Owl in the Pacific Northwest (Noon and McKelvey 1996b). Population viability analysis models also benefit from incorporating uncertainty directly, using Bayesian techniques (Goodman 2002, Taylor et al. 2002, Wade 2002) or exploring the role of estimation error and uncertainty on model outcomes (Parysow and Tazik 2002). They also benefit from using an explicit decision analysis framework (Possingham et al. 2002, Drechsler and Burgman 2004). Ralls et al. (2002) provide an important set of guidelines for using PVA in conservation decisions.

The additional complexity of stochastic single-population models compared with deterministic matrix population models is an example of a tradeoff that we will revisit frequently. On the one hand, stochastic models are more appealing to field biologists because they are more realistic and include more processes; therefore, they can incorporate more of what field biologists know or hypothesize about the system. On the other hand, not only are estimates of demographic rates required, as in matrix population models, but also mechanistic relationships between the environment and demography (e.g. rates and intensities of catastrophes, long-term trends in carrying capacity, and forms of density dependence) must be documented. Finally, the accuracy with which demographic rates are depicted should be determined through model-validation studies.

Even with this added complexity, stochastic single-population models do not consider several factors that commonly occur in nature and may affect extinction rates. Immigration and dispersal are rarely incorporated in single-population models but can have important effects on population dynamics (Stacey and Taper 1992). Individuals may often be found not in a single panmictic population, but rather in spatially distinct subpopulations distributed widely across the landscape and connected by infrequent dispersal events. In addition, demographic rates and their variances may vary spatially among different habitats or subpopulations (Pulliam 1988). Typically, other types of models are used to address these complexities, namely metapopulation models and spatially explicit models.

Metapopulation Models

Metapopulation models have become important tools for understanding the relationships between landscape structure and population dynamics. By incorporating spatial characteristics of local populations, metapopulation models provide a means of analyzing and predicting the response of individual species to habitat fragmentation and other features of landscape structure. They also provide a degree of realism that appeals to managers and researchers.

Whether a metapopulation or single-population model is the more appropriate way to describe population dynamics depends on what is known, or believed, about population structure and the importance of landscape-dependent processes. Faced with the need to model a species living in a patchy landscape, it is important to determine whether the additional processes that metapopulation models incorporate are important and necessary to depict population dynamics accurately with respect to the objective of the modeling exercise (Hanski 2002, Harrison and Ray 2002). A key question is how fragmented the system is relative to the dispersal ability of the species being modeled. Several types of evidence can be suggestive of metapopulation dynamics. Turnover, or the extinction and recolonization of local populations, is the most widely cited (Hanski 1992) but is not a necessary condition for a population to be treated as a metapopulation (Kareiva et al. 1997, Hanski and Gaggiotti 2004). Unfortunately, long-term data on turnover are rarely available, and turnover may not be observed in metapopulations that are driven more by "rescue" than by recolonization (Stacey et al. 1997). Rigorous approaches to investigating processes that produce metapopulation dynamics are reviewed by Ims and Yoccoz (1997) and exemplified for various taxa in Hanski and Gaggiotti (2004). Techniques employed include radiotelemetry, and enclosures and experimental releases to elucidate the role of dispersal and emigration in population dynamics. Modern mark–recapture methods, especially multistrata models, are becoming increasingly important for estimating movement and demographic parameters in marked populations (White et al. 2002). Robust statistical techniques recently have been developed to estimate patch occupancy while accounting for imperfect detection rates and missing data (MacKenzie et al. 2002, 2003, 2006). A complicating factor for metapopulation modeling is whether the focal species inhabits a highly dynamic landscape or a "shifting mosaic" of patches, because most metapopulation models assume that the configuration and quality of habitat patches are relatively static (Wiens 1997; but see Hastings 2003, Hanski and Gaggiotti 2004). For example, Fleishman et al. (2002) found that occupancy and turnover were better modeled by measures of habitat quality than by patch area or isolation. When present, strong temporal changes may dominate the dynamics of a system, and accounting for such changes in habitat quality in a metapopulation model may be difficult (Thomas and Hanski 2004). Even if a landscape is nearly static, the population dynamics within a system may not fit a metapopulation framework. For example, if a system is functioning as a single population or as a collection of independent populations with no migration, use of a metapopulation model may be inappropriate (Harrison 1994, Harrison and Bruna 1999).

Recent years have seen a proliferation of metapopulation models. Below, we outline some of the characteristics of these models and discuss the merits of different models from a conservation perspective.

Structure of Metapopulation Models

Metapopulation models can be grouped into three classes: theoretical, occupancy, and patch models. Theoretical models (Hanski and Gaggiotti 2004) typically make simplifying assumptions about the spatial characteristics of patches—for example, assuming equal patch size or uniform distribution of patches (Lamberson et al. 1992, Gilpin 1996). They may use "neutral landscapes" (With 2004), lattice models (Hanski and Gaggiotti 2004), or method of moments or other mathematical approaches (Bolker 2004). These models typically do not use site-specific information; consequently, their purpose is to provide general insights into metapopulation processes. These models tend to be very general in nature, and they have made fundamental contributions to our understanding of metapopulation dynamics.

Occupancy models have been developed and popularized by Hanski (1994a, b), who used incidence functions to model metapopulations. Hanski's approach has been modified and elaborated into a class of models commonly referred to as "stochastic patch occupancy models" (SPOMs; Hanski and Gaggiotti 2004). Etienne et al. (2004) provide an overview of different types of SPOMs. The incidence function approach combines the pattern of patch occupancy with relationships derived from biogeography to estimate colonization and extinction rates (Hanski 1999), whereas the patch-turnover approach is based on documenting patch colonizations and extinctions (Sjogren-Gulve and Ray 1996, Kindvall 2000, Sjogren-Gulve and Hanski 2000). A key characteristic of SPOMs is their reliance on simple presence–absence data obtained from single or multiple surveys to estimate the probability that a patch is occupied and to model metapopulation dynamics (Hanski 1998, Thomas and Hanski 2004). Morris and Doak (2002) provided computer algorithms for developing occupancy models using maximum-likelihood (ML) techniques or logistic regression, and Moilanen (2004) and Grimm et al. (2004) have created generic software to develop occupancy models. Grimm et al. (2004) also provide an example for Capercaillie (*Tetrao urogallus*). An important recent advance for parameterizing occupancy models was made by MacKenzie et al. (2002, 2003). They extended mark–recapture statistical approaches to estimate patch occupancy, colonization, and extinction rates that account for imperfect detection. These techniques have been implemented in the programs MARK and PRESENCE (see Acknowledgments). Once parameterized from field data, an occupancy model can then be applied to other metapopulations of the same species or used to simulate metapopulation dynamics by substituting new patch-area and patch-isolation data in a simulation model (Hanski 1999).

A rapidly growing list of species has been studied using occupancy models, but bird studies are notably rare. Hanski and Gaggiotti (2004) searched literature published from 1996 to 2000 and found that birds were the least-studied taxa, compared with mammals, fish, butterflies, and plants. Avian examples include Neotropical migrants (Villard et al. 1992), Pacific Island birds (Cook and Hanski 1995), and the European Nuthatch (*Sitta europaea*; ter Braak et al. 1998). The paucity of patch-occupancy models for bird studies may reflect the relative scarcity of unoccupied suitable habitat or turnover, perhaps in part because of the greater dispersal ability and longevity of birds as compared with other organisms.

Patch models are perhaps the most widely used metapopulation models (Fig. 3B). In patch-based models, the smallest unit in which population size is tracked is the individual habitat patch. Individuals reside only within these patches and may move or disperse among them, but they are not permitted to survive in the surrounding landscape. Thus, metapopulation models incorporate spatial relations in a rather restricted manner, because spatial data are limited to a matrix of interpatch distances and patch sizes. Neither the location of individuals within a patch nor the types of landscape matrix that surround a patch are included; however, both may affect survival or reproduction and are often included in spatially explicit models (Fig. 3C, D). Population dynamics within each patch are usually modeled as in stochastic single-population models, with the additional steps of determining the numbers of individuals that will disperse from and migrate to each patch. Patch models are popular because easy-to-use software packages are readily available (Lindenmayer et al. 1995).

The choice of model type may depend, in part, on the availability of data needed to parameterize the model. Data requirements for metapopulation models vary considerably, ranging from a few parameters for some theoretical models to dozens of parameters for complex patch models (Beissinger and Westphal 1998, Grimm et al. 2004, Moilanen 2004). Theoretical models typically have fewer input variables, requiring only very coarse, general data. For example, Levins's (1969) model that introduced the metapopulation concept had only two key parameters, extinction and colonization rates, which

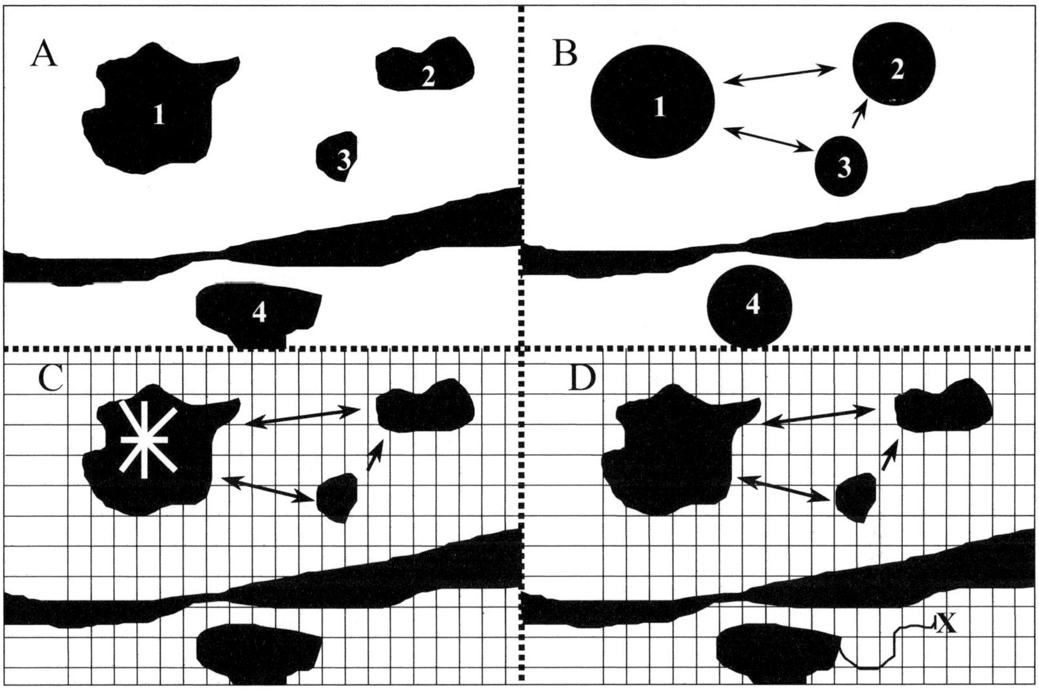

FIG. 3. Demographic PVA models with various degrees of spatial explicitness, modified from Gilpin (1996) and Beissinger and Wesphal (1998). (A) The landscape includes four patches of forest (shaded gray and numbered 1–4) surrounded by a matrix of agricultural lands (shaded white) and dissected by a river (shaded black) that acts as a barrier to dispersal. (B) Patch metapopulation model for the landscape in (A) showing linkages between patches by dispersal indicated by arrows. Each patch has its own internal population dynamics. (C) Spatially explicit cellular automata or grid model for the landscape in (A). The landscape is now represented by grid cells, each with its own population size, immigration and emigration rate, and fecundity and survival rate based on the grid-habitat characteristics and the habitat characteristics of surrounding grid cells. Occupied cells have stars, dashed lines indicate influences of neighboring cells, and arrows show potential dispersal among patches. (D) Individually based model where the movement path, mortality (indicated by an "X"), and reproduction of each animal (star) is tracked across the landscape. Survival and fecundity may differ between the matrix and the patches, as well as within the patches in relation to distance from the edge.

predicted the fraction of occupied patches. Occupancy models have a fairly standardized set of input parameters (Morris and Doak 2002, Grimm et al. 2004). These include simple field data consisting of patch area, occupancy, interpatch distances, and several parameters that are typically estimated using nonlinear ML regression or related techniques. Examples of the latter include parameters expressing the effect of distance on migration rate or the relationship between emigration rate and population density. For example, a model by Hanski (1994b) included eight parameters: carrying capacity, density-independent growth rate, strength of density dependence, demographic stochasticity, environmental stochasticity, emigration rate, migration mortality rate, and interpatch distances. Some of these parameters require information that is nearly impossible to obtain in the field (e.g. rate of mortality during migration), but techniques that require a few simplifying assumptions are available to estimate them from readily obtainable field data (Hanski 1994b, 1999). Accurate estimates of some parameters may be unimportant to overall model accuracy (Hanski 1994a), but Ims and Yoccoz (1997) suggest that precise and unbiased estimates of emigration, migration, and colonization rates may be especially important for systems with high rates of transfer among patches. This argues for the importance of applying statistical techniques that incorporate uncertainty in detection when estimating movement and occupancy rates (White et al. 2002,

MacKenzie et al. 2003). Sensitivity analysis and comparative simulation experiments can also be used to evaluate model uncertainty (Grimm et al. 2004).

Patch models typically require more detailed data than theoretical or occupancy models (Akçakaya 2000a, b; Hanski and Gaggiotti 2004). Simple patch models may require patch carrying capacities, population growth rates, and immigration and dispersal rates among patches. More complex models may have many input requirements in common with the matrix or stochastic single-population models discussed above. For example, age- or stage-structured vital rates may be required, with the added twist that these parameters may vary among patches.

Patch models also require parameters related to spatial structure. These include whether rates of dispersal and immigration are dependent on density within a patch and, most importantly, the degree of correlation among patches for environmental stochasticity (Lindenmayer et al. 1995, Beissinger and Westphal 1998). Environmental stochasticity is typically assumed to be perfectly correlated among patches, meaning that all patches simultaneously experience the same good or bad conditions. This assumption reduces the influence of processes important to metapopulation dynamics, such as the rescue effect (Morris and Doak 2002). If vital rates within patches vary synchronously and if vital rates do not vary greatly among patches, the metapopulation behaves much like a single population; but if they vary asynchronously, the dynamics of a metapopulation can be very different from those of a single population (Stacey et al. 1997). Such variables are extremely difficult to measure in the field, yet they may have a major effect on model results and population behavior. Sensitivity analyses are especially important when using models with poorly known parameters (Beissinger and Westphal 1998, Mills and Lindberg 2002).

Outputs from the various metapopulation models differ greatly and may also be a determining factor in the choice of model type. Because theoretical models may differ greatly in their underlying assumptions, few generalizations can be made about their output. Occupancy models usually yield estimates for turnover rates and the equilibrium proportion of patches that are occupied. Grimm et al. (2004) recommend "intrinsic mean time to extinction" (Grimm and Wissel 2004) as the primary output from their generic model. Once an adequate occupancy model is created for a species, it can be applied to other landscapes to make predictions about the equilibrium occupancy rate for new sets of patches (Hanski 1999, Hokit et al. 1999). Also, by changing the number or configuration of patches, the relative importance of each patch to metapopulation persistence can be evaluated (Hanski 1994a, 1999). Thus, these models can be used to evaluate the suitability of a landscape for a species and the conservation value of particular habitat patches. Because they are based on presence–absence data and are designed to predict patch occupancies, occupancy models generally are not used to predict population densities or actual population size. Patch models have an advantage here; in addition to turnover rates and equilibrium occupancy, they also provide population trajectories, risk of population extinction or decline, and related measures of population size (Akçakaya and Atwood 1997, Hokit et al. 1999).

Estimates for parameters, such as dispersal or the correlation of vital rates among patches, are often based on sparse data. Also, there is often little basis for choosing among different structures for modeling dynamic processes, such as conditions that cause individuals to migrate from one patch to another. Sensitivity analysis can reveal how important a parameter or assumption is to the output of a model; this indicates how much the reliability of model output is compromised by difficult-to-estimate parameters or by the particular way a process is modeled (Beissinger and Westphal 1998, Grimm et al. 2004). With metapopulation models, sensitivity analysis is used primarily for this purpose rather than to determine the best parameters to alter through management efforts; the latter is done with matrix population models. For example, Akçakaya and Atwood (1997) found large differences in sensitivity among 17 parameters in their patch-based metapopulation model of the California Gnatcatcher (*Polioptila californica*). Such findings may provide assurance that some poorly known parameters are relatively unimportant compared with other, better-known ones. This allows field biologists to focus on measuring the parameters that matter most (Dunning et al. 1995).

Compared with stochastic single-population models, metapopulation models increase demands on field biologists for data to parameterize and validate (Beissinger and Westphal 1998). The ability to validate metapopulation models by comparing simulated and observed data varies among model types. Theoretical models are rarely intended to mimic a particular system accurately; therefore, comparing model output with actual field data usually serves little purpose. Occupancy and patch models usually are applied to particular situations, making validation a possibility. Occupancy models are often validated by using a model developed for one metapopulation to make patch-occupancy predictions for another metapopulation, which are then tested against the actual patch-occupancy data (Hanski 1994a, Hokit et al. 1999). Patch models can also be validated by this approach and by comparing predicted population trends with actual population trends. For example, Wooton and Bell (1992) found a good fit between the population increase of Peregrine Falcons predicted by a model and the observed increase over an eight-year period.

Usually, however, managers and researchers are interested in predicting the likelihood of extinction too far into the future for validation to be feasible. Also, field situations that provide true replicates to develop a probability of extinction are exceedingly rare (but see Hanski [1999] for possible examples). Data sets with trajectory or extinction information are often too limited in spatial or temporal extent to compare with model output, making strong statistical validation of these models difficult. One approach is to develop different models of the same system independently and then compare the models for concordant output (Noon and McKelvey 1996b, Kindvall 2000, Hokit et al. 2001). Another approach is to test specific components or assumptions of the model (Thomas et al. 1990). Such tests are valuable, but they do not demonstrate the model's overall accuracy in predicting the future. Consequently, many researchers discourage reliance on predictions projected far into the future from these simulation models. Instead, they recommend using the models for other purposes. These include identifying variables that are likely to be most important, elucidating system dynamics, generating new hypotheses, and evaluating the relative effectiveness of different management approaches (Akçakaya and Atwood 1997, Beissinger and Westphal 1998, Reed et al. 2002, Grimm et al. 2004).

Use of Metapopulation Models in Conservation

Metapopulation models differ significantly in their potential for use in conservation. Theoretical metapopulation models are usually designed to derive a better understanding of general conservation principles about how metapopulations work (Hess 1996a, b; With 1997, 2005; With and King 2001), though they can be designed to model specific situations. Occupancy models are highly practical, because they directly incorporate site-specific field data; they can make short-term predictions about the patch occupancy of a given species for different configurations of patches, imagined or real. Thus, occupancy models can be useful in managing landscape structure. An important criticism of occupancy models is that they assume "quasistationarity"—in other words, that the data used to parameterize the model came from a system with no increasing or decreasing trend in patch occupancy (Hokit et al. 2001, Hanski 2002). Many present-day landscapes may be far from equilibrium because of recent human-induced fragmentation. However, recent studies suggest that occupancy models may be fairly robust to violations of this assumption (Hanski 1999, Hanski and Gaggiotti 2004), and modifications to occupancy models have relaxed this assumption (Moilanen 1999, Etienne et al. 2004).

Patch models are less vulnerable to this problem, because they rely on measurements of vital rates that do not necessarily assume equilibrium conditions (Hokit et al. 2001, Morris and Doak 2002). For example, a metapopulation model was developed for the California Gnatcatcher in a rapidly changing landscape in Orange County, California (Akçakaya and Atwood 1997). The model projected a rapid population decline and a high risk of extinction in most simulations. An important implication was that the results were strongly divergent, depending on the lengths of time used in the simulations. Simulations identified a critical time horizon of 30–40 years, at which predicted extinction rates were greatest. Wooton and Bell (1992) used a patch model of the Peregrine Falcon to identify the northern population as a source and the more critically endangered southern population as a sink. They concluded that intensive recovery actions on the southern population were probably misplaced, and that the same effort directed at the source population in the north would eventually strengthen the southern population through the source–sink dynamics linking the two populations. Because patch models partly rely on underlying biological mechanisms, they may be applicable to more widely different situations than occupancy models (Hokit et al. 2001). However, estimating demographic rates and numerous other parameters needed for these models greatly increases the demand on field biologists for demographic data, and increased model complexity also brings greater risk of error propagation caused by poor parameter-estimation (Conroy et al. 1995, Beissinger and Westphal 1998, Morris and Doak 2002).

Spatially Explicit Models

An especially complex type of simulation model, spatially explicit models incorporate exact spatial and temporal locations of objects. Objects might include patches of habitat, individual organisms, different populations, or barriers to dispersal (Fig. 3). Spatially explicit models have been used to study a variety of large-scale ecological patterns and processes, such as determining the distribution of suitable habitat at multiple scales (Hatten and Paradzick 2003) and predicting regions of future human–wildlife conflicts (Treves et al. 2004). Their primary use in ornithology has been in the study of regional population dynamics. We concentrate on spatially explicit population models (SEPMs) in this review. Spatially explicit population models may be built for entire metapopulations or for single populations, depending on what objects are included and how they are distributed.

Some models are considered "spatially implicit" if they include measures of landscape composition, such as proportions of different habitats in a landscape (Dunning et al. 1992, Donovan et al. 1995, McGrath et al. 2003). For example, the metapopulation models reviewed above are spatially implicit, because they include some spatial information, such as patch size and distances between patches. However, those models are not completely "spatially explicit," because they operate without tracking specific locations of individuals within each habitat patch or kinds of habitat surrounding the patch, each of which may affect demography.

Spatially explicit population models often include individuals as objects, which are placed in known locations in space and assigned habitat-specific demographic traits based on these locations (Fig. 3). The models then total the results of all individuals to obtain population statistics, such as population size at a specific time, population trajectories, or time to extinction. Thus, SEPMs are often individual-based models (IBMs), but their main outputs are population characteristics and processes. Spatially explicit models can be used to model movements of individuals across a diverse landscape (Cooper et al. 2002) or responses of populations to changing landscape structure (Akçakaya 2000b).

Individual-based models have several advantages over the aggregate population models discussed above. They can describe population traits with distributions rather than mean values, explicitly represent individual performance, incorporate local interactions, and yield a mechanistic rather than descriptive approach to modeling (DeAngelis and Rose 1992). Individual-based models appeal to field biologists, because they simulate important processes in the lives of individual organisms, the scale at which population dynamics are determined. Thus, IBMs may be employed as population models without a spatial component, as, for example, in modeling larval recruitment in fisheries, where such models have been employed with great success (DeAngelis et al. 1993, Rice et al. 1993). However, IBMs require extensive data on behavior and its demographic consequences. Habitat-based IBMs, for example, require that variation in the expected values of birth, death, and movement rates be explicitly related to habitat variation. Spatially explicit IBMs are the most complex means of representing population dynamics and are most appealing for species that have complicated life histories and inhabit heterogeneous landscapes (DeAngelis and Rose 1992, Murdoch et al. 1992, Walters et al. 2002).

Structure of Spatially Explicit Models

Spatially explicit population models typically take one of two forms (Fig. 3C, D). Grid- or cell-based models track population sizes in cells that are typically the building blocks of larger habitat patches. Cells do not have to be equal in size (Walters et al. 2002). Cells are influenced by the inputs and outputs of neighboring cells. Grid-based models can be used to monitor separate populations in each cell, especially when the modeled populations are so large that following each individual may be impracticable. Such population-based models are commonly used for abundant organisms such as plants, insects, or rodents (Bradstock et al. 1996, Price and Gilpin 1996). McCarthy et al. (2000) presented an avian example of a population-based model. They constructed a spatial PVA model of two Australian treecreepers (family Climacteridae) in 39 remnant patches of native habitat. The model simulated population dynamics in each occupied patch, and allowed dispersers to link between-patch dynamics. Most avian SEPMs are individual-based models that track the location of every individual and calculate population sizes from counts of individuals in each location. In these models, population dynamics are the outcome of processes that occur at the level of the individual. Individual-based models have been used to simulate responses to regional management practices (Boone and Hunter 1996), to evaluate translocation options (Akçakaya et al. 1995), and to simulate the effects of forest management policies (McKelvey et al. 1993, Lamberson et al. 1994, Liu et al. 1995, Walters et al. 2002).

The crucial part of spatially explicit modeling is linking a population model to a landscape map (Akçakaya 2000a). The map can be as simple as a set of patches, some defined as suitable habitat and others as unsuitable for a particular species. More complex maps can include habitat variables that contribute to classifying habitat suitability for each location. All models, however, require a mapping of the locations of good and poor habitat from the standpoint of the species considered. The development of extensive Geographic Information System (GIS) databases for conservation has allowed the incorporation of real-world landscapes into spatial models. When a GIS-database includes the distribution of habitat patches, a landscape grid can be created by overlaying a grid of cells on top of the original map and then assigning habitat characteristics from the original map to each cell. This works most easily if the grid cell size is considerably smaller than the polygons of the original map.

The population model that is linked to the landscape map can take different forms, depending on the simulations being done. A typical format is a life-history simulator that classifies each individual in the model by sex and age and then takes it through an annual cycle of breeding, survival during the nonbreeding season, and dispersal. Ideally, individuals are assigned habitat-specific demographic traits, depending on where they are located on the landscape grid. Their life history may also depend on location. For example, the likelihood of successful dispersal may depend on proximity to vacant areas of suitable habitat. In models that treat the sexes separately, reproductive success may be related to the spatial distribution of individuals of the opposite sex. More complex social and reproductive strategies, such as obligate flocking and cooperative breeding, have recently been incorporated in SEPMs (Cooper et al. 2002, Walters et al. 2002). In all these cases, the model sums across individuals and landscape patches to calculate population characteristics at various time steps.

Most published SEPMs for bird species are developed for species that are rare, endangered, or of strong management interest. Typically these species are not migratory. These species characteristics are partly attributable to the extensive data needs of SEPMs. Endangered species programs are more likely than other research groups to have the necessary funds and logistical support to collect such data.

Spatially explicit population models are extremely data-hungry. Ideally, they require extensive data on the distributions of habitat types and quality across landscapes, habitat-specific demography from the landscape, and an accurate idea of dispersal patterns and movement rules. Such detail is rarely available. Some of the needed data can be gathered relatively easily; for instance, existing GIS databases can provide the habitat distributions. However, the maps generated from GIS systems often need to be verified by visiting the areas. Maps of habitat distributions are useless without ecological knowledge of how habitat quality affects reproduction and survival. Long-term survey data can provide estimates of habitat-specific density, but locally accurate demographic and dispersal data are generally not available for most species. To provide the data these models require, field biologists must partition their demographic data by habitat quality (Krementz and Christie 1999) and acquire extensive data on dispersal and movement patterns.

Like most models, SEPMs conform to the "garbage in, garbage out" rule. The reliability of the model depends on realism of structure, validity of the coding algorithms, and accuracy of the data used to initialize the simulations. In building a model of a real-world landscape, the physical layout of the landscape can be confirmed with remote sensing data, and thus is usually not the limiting factor in the model. Larger problems are usually presented by attempting to connect habitat suitabilities to various patch types, parameterizing the population model, and building realistic dispersal rules. For each of these steps, there is no substitute for accurate natural-history information for the region and species under study.

Ruckelshaus et al. (1997) constructed an SEPM of hypothetical populations to determine the consequences of parameter error in general model performance. Large errors in model predictions resulted from inaccurate parameterization of dispersal characteristics. Errors in landscape classification had smaller effects on model performance. Wennergren et al. (1995) also found that model results were highly influenced by error associated with dispersal parameters. If these results are correct, they suggest caution in widespread use of SEPMs, because the models were most sensitive to errors in parameters that are typically the most difficult to estimate. Mooij and DeAngelis (1999), however, argue that Ruckelshaus et al. (1997) greatly overestimated the inaccuracies of error propagation in these models. They also showed through a simulation exercise that SEPMs may suffer less from uncertainty in parameter estimates than simpler models (Mooij and DeAngelis 2003) and proposed a strategy of model development to reduce the chance of significant error propagation.

The second main theme apparent from a review of published avian SEPMs is that nearly all models have been developed for permanent resident species. In part, this may reflect the fact that many species of management interest are resident species (e.g. Northern Spotted Owls). But some rare and endangered species are fully migratory (e.g. Whooping Crane and Golden-cheeked Warbler [*Dendroica chrysoparia*]), so the relative lack of SEPMs for migratory species is striking. In concept, the development of a SEPM for a migratory species on its breeding grounds is straightforward, except for the dispersal routines. Migrants show differences from permanent residents in how they disperse to initial breeding sites; these differences present challenges to model developers.

In the life-history simulator in a typical SEPM designed for a permanent resident, individuals (both adult and juvenile) are subject to mortality during the nonbreeding period and then disperse to find suitable breeding locations. Typical models allow surviving adults to be site-faithful to previous breeding locations, and juveniles to inherit natal sites when their parents do not survive or to start dispersing from their natal patch. For permanent resident species, therefore, the number and location of potential dispersing individuals is easily determined for each year of a simulation—they are based on the results of mortality during the preceding nonbreeding period.

For a migratory species, however, these basic components of the dispersal subroutine are not trivial decisions for the model developer. First, field studies of migratory birds have demonstrated that few locally produced offspring return to breed close to their natal territory (Nolan 1978). Thus, each year, the potential new breeders that search for available breeding positions are not local birds and do not have specific locations from which to begin the dispersal phase. The modeler must decide how to place these individuals on the landscape grid at the start of dispersal. Are the new dispersers placed at random into any grid cell or only into suitable-habitat cells? Are they initially placed in the southernmost cells, to imitate a migratory movement from southern wintering grounds? Field studies of how migratory birds make their initial habitat choices provide few guidelines.

Perhaps more important for model results, the model developer must determine how many dispersers to recruit into the population each year. For closed populations of permanent residents, the number of dispersers is determined by survival of local adults and juveniles from the previous year. But this is not necessarily true for migratory species, unless the production and survival of local individuals is correlated with regional populations—that is, if results of the model from the previous year can be used as a guide to regional population dynamics.

Although this may be true in some cases, there are many cases where local demography would not be a good guide. For instance, if the local site being modeled is a sink population, then the number of juveniles produced locally would not be a good measure of regional productivity. If the local population is a consistent source population, then productivity and survival may vary much more outside the local population than within it. Pulliam et al. (1995) demonstrate that, even at the landscape scale, the relative quality of local sites can have strong effects on local dynamics in surrounding regions. Conceptually, there are ways to overcome these problems; but so far, no SEPMs for migratory species have been published. King et al. (2000) developed a spatially explicit model that they used to assess the status of Henslow's Sparrows (*Ammodramus henslowii*) in a highly fragmented landscape. Using the results of modeling the sparrow population on a hypothetical landscape, they were able to make management recommendations for reversing a severe population decline in this species.

Use of Spatially Explicit Models in Conservation

Spatially explicit models can be used in several conservation contexts. Hypothetical populations can be modeled on artificially generated landscapes to determine the likelihood of population response to a general change in landscape structure. For example, Pulliam et al. (1992) used an SEPM of Bachman's Sparrow (*Aimophila aestivalis*) to determine the importance of a rare habitat type, longleaf pine (*Pinus palustris*) forest, in the sparrow's population dynamics. A series of simulations was done, in which the amount of the rare habitat type was varied in a hypothetical landscape, and a population model based on the life-history characteristics of the sparrow was run on these different landscapes. The population responded disproportionately to increases or decreases in longleaf pine forest, which suggests that this habitat played a more important role in population dynamics, compared with other habitat types in the species' range, than its relative rarity would have suggested (Pulliam et al. 1992). Because this rare habitat type contained other species of management interest, most notably the endangered Red-cockaded Woodpecker, the model results were used to bolster management recommendations that this habitat type be protected and increased where possible.

The second use of spatially explicit models is in the study of effects of specific changes in land use on individual species in particular landscapes. Perhaps the unique contribution of SEPMs is the ability to examine how a proposed management action in an actual landscape might affect a given population. To answer this concern, the actual landscape is usually incorporated into the modeling exercise, because general simulations on hypothetical landscapes do not have the necessary specificity.

Spatially explicit models of Bachman's Sparrow have been used to examine the long-term consequences of timber management in southeastern pine forests. Bachman's Sparrows occupy temporary successional stages of managed pine forest and are strongly affected by creation of such habitat through timber-harvest programs. Pulliam et al. (1992) and Liu et al. (1995) used the model to study landscape effects in this poorly dispersing species and to make recommendations concerning specific management plans in a particular forested landscape. One recommendation was to increase suitable habitat for Bachman's Sparrows by altering the habitat structure of 40- to 50-year-old pine forest (which they generally would not use) to mimic that of the mature forests the species prefers. This recommendation was implemented, and Bachman's Sparrows quickly colonized the modified stands (Dunning et al. 2000).

Spatially explicit IBMs have also been developed for several threatened or endangered cooperative breeders, including the Florida Scrub-Jay (*Aphelocoma coerulescens*; Stith et al. 1996) and the Red-cockaded Woodpecker (Letcher et al. 1998, Walters et al. 2002). In both cases, this approach was used because of the spatially restricted dispersal behavior of helpers, which cannot be incorporated into simpler models. The Red-cockaded Woodpecker model was later modified for a study of the effects of landscape change on another cooperative breeder, the Brown Treecreeper (*Climacteris picumnus*) of Australia (Cooper et al. 2002). Spatially explicit population models were also important in making management decisions for the Northern Spotted Owl (McKelvey et al. 1993) and in guiding land-management decisions in southern California that affect the endangered California Gnatcatcher (Akçakaya 2000a).

Spatial models have been influential in the research and management decisions involved in restoring the Florida Everglades (Sklar et al. 2001, 2005). Models have been developed to track the changes in regional hydrology throughout the Everglades as restoration is implemented, as well as changes in associated ecological factors such as the distribution of natural fires within the system (Lockwood et al. 2003). Avian models were used to evaluate how changes in hydrology and fire regimes in the Everglades would affect population dynamics of Cape Sable Seaside Sparrows (*Ammodramus maritimus mirabilis*; Pimm and Bass 2002, Lockwood et al. 2003), Snail Kites (Curnutt et al. 2000, Mooij et al. 2002), and wading birds (Fleming et al. 1994, Curnutt et al. 2000). The models typically ranked alternative management scenarios with a goal of comparing the relative responses. Although no single management alternative is beneficial to all studied species, multispecies modeling allowed managers to better understand the potential tradeoffs in responses of species.

Attempts to validate SEPMs are rare in the literature (Conroy et al. 1995, Schiegg et al. 2005). Unfortunately, as with stochastic single-population and metapopulation models discussed above, many SEPMs used in conservation applications are designed to predict population patterns that operate at too great a spatial or temporal scale to permit direct validation (e.g. extinction rates predicted in 50 or 100 years). One common approach to model validation is to parameterize the model with data from one region of interest and then apply the model to another area to determine if model predictions are robust (Akçakaya and Atwood 1997, Akçakaya 2000a, Schiegg et al. 2005). McCarthy et al. (2000) built a model to see how fragmentation affects two species of Australian treecreeper using data from outside the main study area and then determined the model's ability to predict extinction and colonization events within the region. The model tended to underestimate the observed number of extinctions and colonizations, perhaps because of imperfect detection probabilities derived from field surveys. The validation attempt for the treecreeper model improved model development by highlighting the model components that may have contributed to errors in model predictions (McCarthy et al. 2000). Schiegg et al. (2005) compared results of a model of dynamics developed for one population of Red-cockaded Woodpeckers to the dynamics of a different population. In this case, the model proved to be accurate in predicting most population parameters of the second population, though Schiegg et al. (2005) concluded that better estimates of some aspects of dispersal behavior would further improve the model.

Another approach is to predict short-term patterns that can be tested against field data. In the Bachman's Sparrow model, for instance, Liu (1993) used a randomization procedure to initialize population distributions onto a map simulating part of the Savannah River Site, South Carolina. This procedure was based on field data of habitat use in a single year. Liu (1993) then produced predicted distributions of bird density across different-aged stands in future years. When distributions were documented in subsequent years, the fit between predicted and observed distributions was satisfactory. Less satisfactory results were obtained by matching observed versus predicted distributions of dispersing individuals in an unusual landscape that developed later (Dunning et al. 2000). The latter result suggested that the dispersal subroutines could be improved.

As with metapopulation models, sensitivity analyses have been conducted primarily as simulation exercises to identify key parameters, rather than to gain insights into management. In more sophisticated analyses, several parameter values are varied in a factorial experimental design, allowing the detection of interaction effects among parameters (Pulliam et al. 1995). Sensitivity analyses have been done with the Spotted Owl, Bachman's Sparrow, Red-cockaded Woodpecker, and California Gnatcatcher models. The latter model proved to be most sensitive to density-dependent effects, weather-related catastrophes, and adult survival and fecundity (Akçakaya and Atwood 1997). In the Bachman's Sparrow model, demographic parameters, such as adult survivorship and habitat-specific reproductive success, were identified as having a greater influence on model results than landscape characteristics or dispersal (Pulliam et al. 1992). The Red-cockaded Woodpecker model was also more sensitive to mortality parameters than to dispersal parameters, except under conditions of extremely low population density (Letcher et al. 1998).

The appeal of spatially explicit models lies in their ability to capture the actual variability of specific landscapes, but this also represents one of their greatest weaknesses. If a model is built to test the response of a specific population in a particular landscape, it is not clear whether the results can be generalized to other species in the same landscape, or to the same species in a different location. Different species move over landscapes in different ways (Belisle 2005). A mapping of landscape change at a particular spatial and temporal scale for sparrows may not be relevant for modeling Red-tailed Hawks (*Buteo jamaicensis*). Research to determine the flexibility of spatially explicit modeling across different landscapes and different species is urgently needed; one published example is the application of the Red-cockaded Woodpecker model to an Australian species with a similar life history (Cooper et al. 2002). This need is compounded by the large amounts of natural-history data required to parameterize models accurately for multiple species on various landscapes.

Simulating hypothetical populations on artificially generated landscapes can provide general results that test landscape ecology theory (Fahrig and Merriam 1994, With 2005). For instance, With and King (2001) modeled songbird populations on artificial landscapes to test the relative importance of edge effects and area sensitivity. They determined that the effect of habitat fragmentation could be reduced by retaining some large habitat patches within a landscape. Although the results of models of hypothetical populations can produce management insights, resource managers are reluctant to accept the relevance of theoretical results. With any complex model, the tradeoff between realism and generality is disproportionate. Loss of generality is a cost that users of spatially explicit models sometimes accept in generating results needed for specific conservation settings.

Not all landscape models need to be spatially explicit. The data-demanding nature of these models and the computer expertise required to design them suggest that simpler modeling approaches should be adopted whenever possible. For instance, an important landscape variable affecting some bird populations is the amount of suitable habitat within a reasonable distance of current populations (Pearson 1993, McGrath et al. 2003). The specific location of the suitable habitat may not explain much of the residual variation in population dynamics. This is an effect of landscape composition (Dunning et al. 1992) and can be modeled with a simpler analytical approach (e.g. Olson et al. 2004). Urban (2005) has recently proposed a graph analysis technique that produces results that are similar to spatially explicit metapopulation models but does not require the same level of detail.

An example of how to determine the proper degree of model complexity is provided by McGarigal and McComb's (1995) excellent landscape study of breeding birds in mature forest of the Oregon coastal range. Most bird species showed strong landscape-composition effects, such as increasing their numbers in areas with more old forest, but few species were affected strongly by the spatial arrangement of habitat patches. A landscape model for most species would not need to be spatially explicit, because the model would need only track the area of late-seral habitat in a given region and not the location. Species that responded to edge, patch size, or other aspects of spatial arrangement might be more profitably modeled in a spatially explicit manner.

Several theoretical studies have suggested that the effects of placement of objects in the landscape are most apparent when the critical habitat resources constitute 5–30% of the landscape (Fahrig 1992, Lamberson et al. 1992, Andren 1994). When the critical habitat type makes up <5% of the landscape, the patches may be too isolated to be found easily by dispersers. When the habitat exceeds 30% of the landscape, dispersers may move easily throughout the landscape and few effects of landscape structure are significant. Overall habitat loss has a greater effect on population loss than fragmentation alone (Fahrig 1997), which suggests that landscape composition effects should be examined before developing a fully spatial model for a specific conservation situation.

If one accepts the accuracy of a given model, spatially explicit models allow the study of population dynamics at the spatial and temporal scales at which land-use decisions are made. Because they directly incorporate the habitat-change processes that may affect the population dynamics of many species, SEPMs should be able to provide important insights. But we must understand key biological processes well to incorporate them correctly. Spatially explicit population models, like the other models we review, cannot substitute for carefully designed field studies. Their construction, parameterization, reliability, and validation are critically dependent on field studies. On the other hand, SEPMs can benefit field studies by identifying what parameters most affect population dynamics and what data may be most valuable for conservation (Dunning et al. 1995).

Genetic Models

All models discussed to this point represent descriptions of population dynamics. The current and future dynamics of populations are a major concern in conservation, but they are not the only concerns. Below, we discuss models used to investigate two related questions, genetic variation and geographic distribution. We begin with genetics.

The chief conservation concerns regarding genetic variation are preservation of genetic variability and avoidance of inbreeding depression. Describing genetic variation is useful, not only in this context, but also in assessing various aspects of population structure critical to population models. Therefore, the word "genetics" is used in several distinct ways in discussions of modeling in conservation: molecular genetic techniques are used to describe genetic variation and to diagnose population structure, taxonomy, and other biological features; whereas genetic models are used to diagnose population status and viability and to test hypotheses regarding recovery actions.

Molecular genetic techniques have been widely applied to conservation problems (Hedrick and Miller 1992, Avise and Hamrick 1996, Haig 1998, Webster et al. 2002). For example, mitochondrial DNA, amplified fragment length polymorphisms (AFLPs), single nucleotide polymorphisms (SNPs), random amplified polymorphic DNA (RAPDs), and microsatellites have all been used in conservation studies to identify metapopulation structure so that appropriate population viability analyses could be done. Molecular techniques have also been used to identify mating systems so that social structure and other features could be parameterized appropriately in demographic PVA models (Haig et al. 1993b, 1994b). Similarly, molecular data can be used to identify individuals in a pedigree to properly incorporate population structure in population viability analysis, to conduct pedigree analyses on a verified pedigree, or to identify species or stocks (Haig et al. 1994a, 1995, 1997, 2003; Daniels and Walters 2000; Jones 2002; Blouin 2003; Gautschi et al. 2003; Ballou and Ralls 2004). A great challenge to these efforts is the common problem of how to construct pedigrees in the absence of information pertaining to the relatedness of known individuals (Gautschi et al. 2003, Russello and Amato 2004). Often, people choose assignment tests on the basis of extrapolations of relatedness from known individuals to unknown. However, Wang (2004) proposed a model that used information about groups, instead of individuals, to manage populations. Molecular genetic applications to conservation have been well reviewed (Avise and Hamrick 1996, Haig 1998), so we concentrate on understanding how models are used to evaluate genetic goals.

Genetic goals related to preserving genetic variability and avoiding effects of inbreeding can be used to define population viability. Genetic goals have been much discussed over the past 10–15 years in relation to the importance of genetic, as opposed to demographic, goals in population recovery efforts (Lande 1988a, Schemske et al. 1994). Briefly, one view is that populations should be managed to maximize genetic variability, because this may increase the likelihood of long-term population persistence. The opposing view is that habitat is being destroyed and populations are declining too quickly for managers to be concerned with long-term genetic goals, and that demographic population goals should therefore be paramount. Both genetic and demographic factors are important in assessing population viability, but their relative importance varies depending on the situation (Haig 1998). For example, when the last of the Guam Rails (*Rallus owstoni*) and Micronesian Kingfishers (*Todiramphus cinnamomina cinnamomina*) were brought into captivity, it was critical to identify close relatives prior to establishing a captive breeding program (Haig et al. 1994a, 1995). Once this was determined, it was then important to expand the populations as fast as possible to reduce the risk of extinction brought about by stochastic demographic events.

Structure of Genetic Models

Effective population size.—Critical to avian population biology and conservation is the theoretical model posed by Wright (1931) when he defined "effective population size" (N_e) as the number of individuals in a population that contribute genes to the next generation. Useful for estimating the magnitude of genetic loss over time in a small population, N_e is a description of the effect of genetic drift on a natural population when conditions are not ideal. Characteristics of an ideal population are both genetic and demographic, and include large and stable size, equal genetic contribution of both sexes, no inbreeding, equal family size, and non-overlapping generations. Most organisms violate some or all of these characteristics. Deviations from the ideal population size, or the relationship between N_e and the censused population size (N), are compared to evaluate the status of an actual population (Waples 2002). N_e is generally substantially less than N, often constituting only 10–25% of total population size (Frankham 1995, Waples 2002). Comparison of N_e with N and changes of N_e over time provide a useful yardstick with which to measure progress of a population toward recovery. Furthermore, managers can examine the specific parameters that are affecting changes in N_e as a means of diagnosing factors inhibiting recovery.

The usefulness of N_e as a means of measuring population status, more specifically its relative vulnerability to inbreeding depression and loss of genetic variability through genetic drift, was recognized early in the history of conservation biology. Franklin (1980) and Soulé (1980) suggested that small populations ideally should attain an N_e of 50 individuals over the short term and of 500 individuals over the long term. This "50/500 rule" was quickly incorporated into recovery planning efforts, such as that directed at Red-cockaded Woodpeckers (U.S. Fish and Wildlife Service 1985). This merging of genetic and demographic information was well intended, but choosing specific numeric goals has not served the purpose, because some managers have adopted the specific numeric goals (e.g. 50 or 500) without regard to the structure of their populations. For the Red-cockaded Woodpecker, a genetically effective size of 500 may require at least 509 breeding pairs or a total population size of 1,322 individuals, including nonbreeding helpers (Reed et al. 1988). Later, Lande (1995) elaborated on the 50/500 rule by proposing that a minimum effective size of 5,000 should be attained. More recently, Franklin and Frankham (1998) have argued that an N_e of 500–1,000 is appropriate. Thus, the debate continues.

Unfortunately, choosing impossible or unrealistic goals has weakened the popularity of using N_e as a concept in management. Many avian populations could never practically attain an N_e of 500, and many have never realized an N_e of 5,000. A more appropriate use of the concept is that proposed by Mace and Lande (1991) when they revised the criteria for listing species for the International Union for the Conservation of Nature (IUCN) Red List. They proposed various categories for listing based on estimates of N_e and other factors, such as amount of habitat alteration and chance of catastrophe. Thus, factors affecting N_e estimates are considered, but managers are not held to specific numeric goals. Below, we discuss how managing for improved N_e can be carried out through pedigree and viability models.

Pedigree analyses.—Perhaps some of the least known, most poorly understood, but helpful genetic models for small populations are pedigree analyses (Fig. 4), which are based on the gene-drop model (MacCluer et al. 1986, Mace 1986, Haig and Ballou 2002). Pedigree analyses were developed for assessing management strategies for captive populations, but they have proved helpful for understanding processes in natural populations as well. They are a way of assessing genetic variability so that adverse effects of inbreeding can be avoided.

Gene-drop pedigree analyses are Monte Carlo simulations in which 10,000 iterations of the model represent sampling of an individual's entire genome. Each founder is assigned two unique alleles at the beginning of a simulation (Fig. 4). Using SPARKS software (Lacy 2004), the model "drops" each of the two alleles for each founder through an established pedigree, generation by generation, assuming Mendelian inheritance. Model iterations generate a distribution of probabilities that individuals in the living population share founder alleles in a proportion related to genetic events encountered over generations. Thus, events such as inbreeding or over-representation (referred to as "over-breeding") of certain individuals will result in a disproportionate number of some founders' genes in the living population. Overall, the gene drop provides an estimate of genetic diversity in the living population.

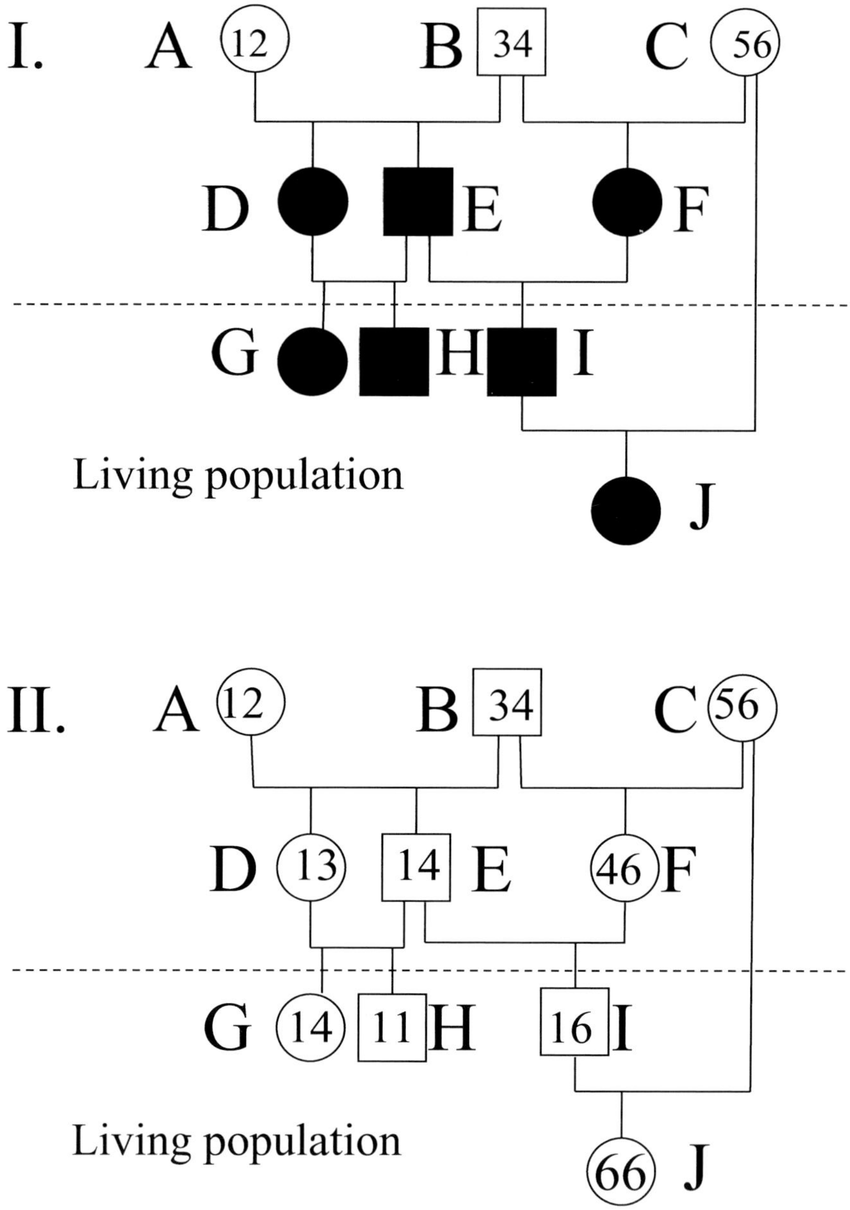

FIG. 4. Gene-drop pedigree model and analysis for a small population. (I) In this example pedigree, each of the three founders (birds A, B, C) has been given different alleles (nos. 1–6). (II) The gene-drop model operates by randomly passing founder alleles through the pedigree from parents to offspring with a 50% chance of each allele's passing to the offspring. After one iteration of the simulation, birds H and J are homozygous, whereas others in the living population (birds G and I) are heterozygous. Thus, 50% of the heterozygosity is retained (two of four living individuals). In the living population, three unique alleles (1, 4, 6) have survived and all three founders are represented. Because some alleles have been lost (2, 3, 5), founder contribution is variable: four of eight alleles in the living population are contributed by founder A, one of eight from founder B, and three of eight from founder C. The simulation would continue by returning to the original starting pedigree (I) and beginning a new iteration of the gene-drop process. This process would be repeated until 10,000 iterations had been completed to produce probability distributions of the model outputs (see text for details).

Assumptions of the gene-drop model include independence among model runs, no linkage, and no selection. The consequences of violating these assumptions depend on the extent to which linkage and selection occur in the natural situation. In a different approach, the model can use a specific pedigree structure, rather than a simulated structure, which adds reality to the exercise. The standard data set required for gene-drop analysis is simple (Table 1), and the model can be applied to either captive or wild populations (Haig and Ballou 2002).

An example of one iteration of the gene-drop model is depicted in Figure 4. In a typical analysis, 10,000 iterations are performed and the results are summarized as the frequency with which neither, one, or both of two alternate founder alleles survive in the living population. From this, the following parameters of the current population can be calculated: founder contribution, the number of unique alleles surviving (allelic diversity), heterozygosity, mean kinship, inbreeding coefficients, and founder genome equivalents. Founder contribution is a measurement of the "presence" of founders in the living population. A founder is counted as present in the living population if either or both of its alleles are present. The gene-drop simulation summarizes the number of times this occurs in relation to other founders. To maximize genetic diversity, all founders must be equally represented in the living population; thus, under- or over-represented founders reduce genetic diversity. The number of unique (i.e. founder) alleles at the end of a gene-drop simulation estimates the representation of founder alleles present in the living population. It differs from founder contribution, in that founder contribution is made if either or both founder alleles are present, whereas number of unique alleles is a direct count of alleles. From a conservation perspective, the goal is to maximize the number of unique alleles in the living population. This is achieved by attempting to equalize founder contribution. The gene-drop approach measures heterozygosity by counting the number of living heterozygous individuals after each run and averaging across runs. Use of founder genome equivalents (FGE) combines maximizing the number of unique alleles and equalizing founder contribution. If 21 founders start a population, there are 42 unique alleles, and FGE is 21 when all founders are equally represented in the living population and the 42 unique alleles are present as well. Previously, population managers strove to maximize FGE. In recent years, the concept of mean kinship has been incorporated into these considerations. Mean kinship is the average kinship between an individual and all others in the pedigree and is used as a measure of genetic importance of an individual in a pedigree. The relationship between FGE and average mean kinship (AMK) is: AMK = 1/2 * FGE (Ballou and Lacy 1995). The manager's goal is to minimize mean kinship or maximize founder genome equivalents so that relatedness among individuals is kept low and inbreeding does not become a problem. Inbreeding is considered by calculating an inbreeding coefficient, which is the probability that two individuals carry alleles that are identical by descent, a measure of the extent of mating of close relatives that managers try to minimize.

The gene-drop model can be verified by comparing simulated values to actual values calculated by hand from the pedigree. However, variables such as heterozygosity cannot be verified using molecular data, because assumptions of the model are not met by measurements of molecular markers. The model begins by assuming 100% heterozygosity among founders (i.e. each founder has two unique alleles), whereas molecular measures take into account the effects of previous genetic events.

The information derived from the gene-drop model describes population structure and

Table 1. Structure of the standard input file ("Me, Ma, Pa" file) used to keep studbooks, to construct pedigrees, and for gene-drop analyses.

Studbook number	Sex	Dam	Sire	Alive (A) or Dead (D)
1	M	WILD	WILD	A
2	F	WILD	WILD	A
3	F	WILD	WILD	D
4	M	2	1	A
5	F	2	1	A
6	F	3	4	A

indicates the importance of different individuals to the living population. The technique can also be used to evaluate various breeding strategies. The consequences of different pairings can be compared by adding hypothetical offspring from selected pairs to the bottom of the pedigree and running the model to see how genetic diversity or population structure is altered. This exercise is particularly useful when setting up programs to introduce or translocate individuals. For example, when Guam Rails were to be introduced on the island of Rota, the gene-drop was used to determine the best pairings to produce chicks for the release (Haig et al. 1990).

One of the most positive aspects of pedigree analyses is that, compared with other approaches, the data sets required are relatively easy to construct, particularly for captive populations, and the programs are easy to run. Furthermore, because the gene-drop model takes into account the exact structure of the pedigree, it is quite realistic. Nevertheless, the greatest obstacle to its use with wild bird populations is the requirement of a deep pedigree derived from marked individuals of molecularly confirmed parentage. For field biologists conducting long-term studies of marked birds, the insight gained into population structure from gene-drop pedigree analysis is a worthwhile dimension that few have explored. For example, when a tiny population of Red-cockaded Woodpeckers had nearly been extirpated from the Savannah River Site in South Carolina, DNA fingerprinting helped prepare a pedigree (Haig et al. 1993a). The gene-drop analysis was then used to measure heterozygosity and compare its loss with the heterozygosity of the original population; to evaluate inbreeding; to determine which founding birds were over- and under-represented in the living population; and, perhaps most importantly, to determine the potential for improving population viability through better breeding practices (i.e. introducing individuals of specific age or sex; Haig et al. 1993b). Subsequent PVAs merged this genetic information with demographic information to provide a more comprehensive plan for recovery.

Use of Genetic Models in Conservation

Population viability analysis models usually take one of the forms of demographic models reviewed above. However, an examination of popular, commercially available population or metapopulation viability software programs indicated that genetic factors, such as reduced fecundity or survival caused by inbreeding depression, were not well incorporated in any of them (Table 2). Among the programs reviewed, VORTEX incorporates genetic factors to the greatest extent, but even with it the genetic data entered can be fairly generic. VORTEX uses the gene-drop model to simulate occurrences of inbreeding and lethal homozygotes, and calculates mean expected heterozygosity and mean inbreeding coefficient at different intervals (Haig and Ballou 2002). Inbreeding depression is incorporated in terms of reduced juvenile survival, and effects of lethal recessives are incorporated through heterozygote advantage. Few studies have calculated lethal equivalents (Ralls et al. 1988), so exact values for lethal equivalents are difficult to find, and the effect of mildly deleterious mutations may not pose a threat to populations with $N_e \geq 25$ (Gilligan et al. 1997). However, the latest edition of VORTEX (version 9.5) allows users to determine the type of mating for the population (e.g. mating by mean kinship). Thus, it uses the pedigrees it creates to make management decisions. A new program called ZOORISK models captive populations

Table 2. Consideration of genetic factors in common population viability models used in conservation planning.

Software program	Model input[a]	Potential to model genetic factors: Inbreeding	Heterozygosity	Genetic distance	N_e	Reference
VORTEX	Individual	Yes	Yes	No	No	Lacy et al. (2005)
GAPPS	Individual	Yes	No	No	No	Downer (1993)
RAMAS	Matrix	Yes	No	No	No	Akçakaya (2004)
INMAT	Matrix	Yes	No	No	No	Mills and Smouse (1994)

[a] Individual = identifies and tracks each individual throughout the simulation; matrix = based on stages or ages in a population.

using PVA simulations and gives them IUCN-like categories for risk, depending on the results (Earnhardt et al. 2004).

Incorporating additional genetic information into PVA models will provide a more complete and realistic evaluation of population persistence, but it will not be simple. Population viability analysis models are demographic models. Thus, genetic factors must be linked to demography. We could begin by trying to incorporate inbreeding into a model. However, one can never tell where or how inbreeding will strike in a population. It could affect life span, fecundity, sperm count, juvenile mortality, or other attributes. Therefore, it would be difficult to attribute the effects of inbreeding to a specific demographic factor. Conversely, there are some generic rules that could be incorporated into models so that they would start to consider genetic factors. For example, a model could assume that for every 20% loss of fitness there would be a 20% increase in inbreeding depression. Another approach would be to use measures such as genetic distance and gene flow to define the relationship among populations in a metapopulation. Currently, some models assume equal distance among populations, whereas others require stating a migration rate, which is often based on an unreliable estimate. Finally, PVA models could evaluate genetic factors independently of demographic factors, such as changes over time in heterozygosity or inbreeding that were estimated by molecular data; they could report a separate probability of persistence based on genetic factors. This probability could be monitored over time or tested against various management options. As in PVAs (Beissinger and Westphal 1998, Reed et al. 2002), the specific probabilities reported will not be as informative as comparisons of changes in persistence values.

The most comprehensive applications of genetic considerations in conservation should use all the analyses described above with the goal of improving N_e (i.e. use molecular tools to determine population structure, perform pedigree analysis, and incorporate genetic factors in PVA). Such an approach has rarely been undertaken.

A more prominent role for genetic models and genetics in conservation depends on resolution of several issues. First, resistance to considering genetic factors needs to be overcome. Toward this end, workers involved in molecular genetic analyses and modeling have a responsibility to translate their results into a format easily understood by those involved in conservation. Second, molecular tools should be more widely employed to parameterize population structure in models used for population-viability or metapopulation-viability analyses. To accomplish this, models with appropriate structure are needed. Finally, because the concept of effective population size is important and helpful to managers, we need to consider it more carefully and more often in population recovery efforts.

Species Distribution Models

The final category of models we will discuss, species distribution models (SDM), are powerful tools for converting individual point-locality data (e.g. museum collection records) into the hypothetical distributional range of a species (Corsi et al. 2000, da Fonseca et al. 2000) or predicted ranges following environmental change (Polansky et al. 2000, Peterson et al. 2002a). Thus, SDMs have great potential utility for avian conservation, especially because avian biologists are often pressed to make recommendations about conserving biodiversity in parts of the world with limited data on species distributions (da Fonseca et al. 2000, Peterson et al. 2000). Compared with other vertebrate groups, however, the large amounts of data available on bird distributions and natural history make them one of the best groups for SDMs (e.g. Edwards et al. 1996).

Over the past 50 years, a wide variety of techniques have been used to link animal distributions to habitat features (reviewed in Stauffer 2002). Species distribution models vary from simple sets of rules based on overlays of environmental and species-occurrence data (e.g. BIOCLIM and simple overlay; Nix 1986, Busby 1991, Loiselle et al. 2003) to sophisticated multivariate analyses (Pereira and Itami 1991, Carpenter et al. 1993), or artificial-intelligence techniques using rule-based sets of algorithms (e.g. GARP; Stockwell and Peters 1999, Godown and Peterson 2000, Anderson et al. 2003). Predicted species distribution maps are also an integral part of gap analysis, which is concerned with locating concentrations of high biodiversity and identifying species, usually terrestrial vertebrates, that are poorly represented by the current set of protected and managed areas (Scott et al. 1993, 1996). The goals of using these models are (1) to establish relationships between habitats and biodiversity; and (2) to establish links between habitat conditions and the viability of one or several species, allowing predictions about habitat changes (Van Horn 2002). Typically, the focus in most ornithological studies has been on modeling terrestrial species.

Traditionally, range maps depicting the distribution of a particular species were either a dot map of specific locations where a species had been recorded or a solidly colored range map enclosing all know sightings (Butterfield et al. 1994). The boundaries of the species range were defined either by connecting specimen or sightings locations (e.g. minimum convex polygons) or, more commonly, by drawing arbitrary lines to encompass the suspected range of the species, like range maps that one might find in a field guide. However, most species do not occupy all areas within their range; rather, the distribution is patchy. Species distribution models attempt to increase the accuracy of range maps by eliminating those areas within the known range where the species does not actually occur.

Structure of Species Distribution Models

Two distinct approaches are commonly used to model species distributions. One "bottom-up" approach is to start with known locations, or point-location data, and then extrapolate that information to the entire range of the species or to an area of interest. O'Connor (2002) called these "forecasting models." For example, the program DOMAIN (Carpenter et al. 1993) takes known distribution points for species and uses map layers of environmental factors, such as climate, soil, and land use, to construct an environmental habitat envelope or "domain" for those points. The envelope is then compared with environmental data for the region under study and a map is produced of similarity ranges of the species' primary domain.

Similarly, genetic algorithm for rule-set prediction (GARP) models include several algorithms in an iterative, artificial-intelligence-based approach (Stockwell and Noble 1992, Stockwell 1999, Stockwell and Peters 1999) using point-source data (e.g. museum specimens). As such, these models use only data on known occurrence, even if data exist on areas of nonoccurrence (Anderson et al. 2003). Usually, occurrence points are divided randomly and evenly within GARP into training and testing data sets (Peterson and Kluza 2003), though Anderson et al. (2003) suggested a protocol for model selection based on omission and commission rates that allows all data to be used in an analysis. The genetic algorithm for rule-set prediction works in an iterative process of rule selection, evaluation, testing, and incorporation or rejection; a method is chosen from a set of possibilities (e.g. logistic regression or bioclimatic rules) and then applied to the training data until a rule is developed or evolved. Predictive accuracy is evaluated on the basis of points resampled from the test data and points sampled randomly from the study region as a whole, which are summarized in a rule-significance measure. Rules may evolve by a number of means that mimic DNA evolution (e.g. point mutations, deletions, crossing over). The change in predictive accuracy from one iteration to the next is used to evaluate whether a particular rule should be incorporated into the model, and the algorithm runs either for 1,000 iterations or until convergence.

A second "top-down" approach is to start with the known distribution of a species and attempt to eliminate those parts of the distribution where the species does not occur. O'Connor (2002) called this approach "exploratory models." Gap analysis uses such an approach (Scott et al. 1993, 1996; Jennings 2000). Landcover maps of actual vegetation are used as surrogates to predict distributions of organisms on the basis of habitat-relationship models. Species distributions are then overlaid on maps of management, ownership, or both, using a geographic information system (GIS). Species with significant amounts of habitat outside the protective envelope of managed lands are defined as "gap species," and unmanaged habitats supporting those species are potential "gaps" in need of conservation planning (Scott et al. 1993). Although gap analysis can be conducted on smaller scales (White et al. 1997, Karl et al. 2000), it is generally considered a "coarse-filter" approach to conserving biodiversity, because it works best at large scales and with common species. For example, Brooks et al. (2004) and Rodrigues et al. (2004) present the beginnings of a global gap analysis. Because it is enacted at such large spatial scales and analyzes many species, gap analysis is the antithesis of endangered species programs that focus on single species that are rare.

As originally proposed by Scott et al. (1993), the basic components associated with a completed gap analysis include (1) a hierarchically based map of actual vegetation; (2) predicted distribution maps for each breeding terrestrial vertebrate species and, in some cases, other taxa (e.g. wintering birds, butterflies); (3) a land ownership–management status map depicting current stewardship; and, most importantly, (4) a spatially explicit representation of areas of potentially high biodiversity and their relationship to current stewardship.

Habitats used by species of interest are defined by the landcover map. Each species is linked to landcover types using habitat-relationship models (Scott et al. 1993, 1996; Butterfield et al. 1994). Briefly, habitat-relationship models are developed by identifying those cover types in which a species might occur (Morrison et al. 1992). In rare cases, a species might be restricted to only one habitat (i.e. one landcover category). More commonly, species occur in several habitat types and may even occur in all habitat types within a given area. The sum of all habitats in which a species might occur becomes the potential distribution of that species. Incorporating the known distribution of a species within a given area, the predicted distribution of a species then becomes all habitats the species might occupy within its known distribution (Fig. 5). Additional model inputs, such as digital elevation data, location of aquatic habitats (e.g. lakes, streams, wetlands), presence of caves, and size of habitat polygons, have been used in several states to refine the predicted distributions of gap species. Using data from the Washington Gap Analysis project, Smith et al. (1997) demonstrated how avian habitat-relationship models can be successfully combined with landcover information.

As in any predictive model, there are four possible outcomes of an SDM linking species to habitats: (1) presence of a species is predicted accurately; (2) absence is predicted accurately; (3) presence is predicted, but the species is actually absent (error of commission, false positive, or type I error); and (4) absence is predicted, but the species is actually present (error of omission, false negative, or type II error). Statistically valid methods for assessing the accuracy of predicted distributions of animals exist, if those four rates can be estimated. An overall index of the performance of a species distribution model is encapsulated by its kappa value (Fielding and Bell 1997), given by the formula:

$$\frac{[(a + d) - (((a + c)(a + b) + (b + d)(c + d))/N]}{[N - (((a + c)(a + b) + (b + d)(c + d))/N)]}$$

where a = number of times when both model and observations predict occurrence, b = number of

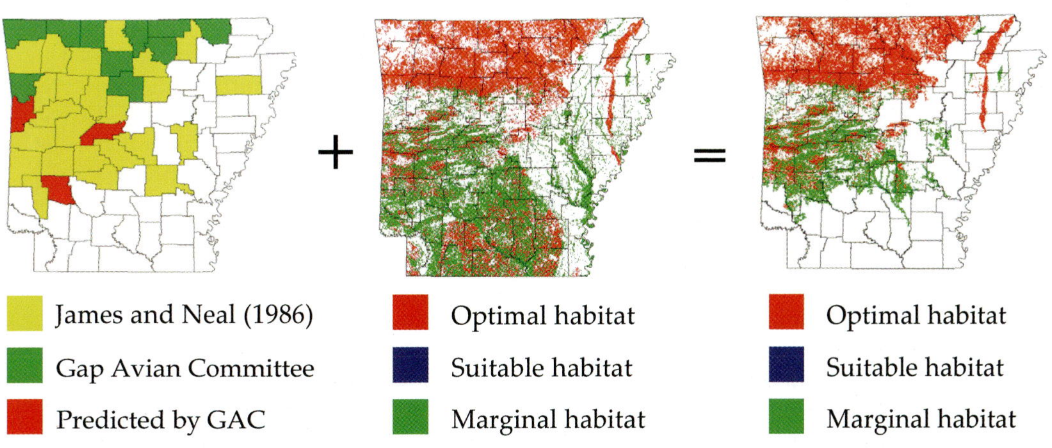

FIG. 5. Example of a gap modeling process for the Ovenbird (*Seiurus aurocapilla*) in Arkansas (see text for details). The known distribution (current range) by county within the state was determined from James and Neal (1986) and a committee of bird experts. The distribution of potential habitats within the state was based on habitats the species is known to occur in based on input from the committee. The predicted distribution is the combination of those two maps, showing the potential habitats within the known distribution.

times when observations indicate absence and model predicts occurrence, c = number of times when observations predict occurrence and model absence, and d = number of times when both model and observations predict absence. These kinds of statistical tests typically require an independent data set that was not used in model development, a requirement that may not be met with most SDMs.

Loiselle et al. (2003) compared the accuracy of 11 modeling approaches to the actual distribution of 11 cotingids of conservation concern in the highly fragmented Atlantic forest of southeastern Brazil using the kappa value. The models they tested included BIOCLIM, simple overlay, three versions each of logistic regression (50%, 85%, and 95% probability), DOMAIN (85%, 90%, and 95% cut-off), and GARP (of five runs, one randomly chosen run [GARP1], cells that were selected in four of five runs [GARP4], and cells that were selected in five of five runs [GARP5]). Results varied dramatically among the various models, with the DOMAIN90 and DOMAIN95 having the highest kappa values.

Loiselle et al. (2003) further cautioned that conservation planners need to evaluate the conservation implications of errors of commission and omission before they start the modeling process. In the case of 11 cotingas, Loiselle et al. (2003) found that models that minimized errors of commission lead to a better selection of reserve networks. In contrast, because animals rarely occupy all suitable habitats within their range, gap analysis models will generally overpredict species occurrence, leading to a higher rate of errors of commission than of omission (Smith and Catanzaro 1996). The best GARP models seem to have low rates of omission and medium to high rates of commission (Anderson et al. 2003). In a direct comparison, gap analysis models had fewer omissions and GARP models had fewer overestimations (Peterson and Kluza 2003).

Individual bird species differ in how accurately they can be modeled, which in many cases can be determined *a priori* (Boone and Krohn 1999). In general, Boone and Krohn (1999) found that sites larger than 1,000 ha with checklists that were based on >10 years of records had the lowest commission errors. In building bird–habitat relationship (BHR) models for 60 species in northern Idaho, Karl et al. (2000) found that increasing model complexity decreased errors of commission more than it increased errors of omission, leading to better overall model accuracy. Thus, it would appear that improvement of avian predictive models should focus on factors that decrease errors of commission.

Evaluating the accuracy of GARP models presents an interesting challenge, because the results vary each time a model is run. This is caused by the way the starting point is selected, typically at random. One approach is to run a model a certain number of times and select areas of habitats on the basis of how often they appear in the model. Loiselle et al. (2003) ran five GARP models and selected a model at random (GARP1), a model based on areas that occurred in four of five models (GARP4), or a model based on areas that occurred in all five models (GARP5). Running GARP models repeatedly and only incorporating those areas that are consistently identified by the models tends to decrease the total area of the final model (K. G. Smith pers. obs.). Anderson et al. (2003) outline a five-step protocol based on accepted levels of omission and commission, rerunning GARP models until a suitable number of models (e.g. n = 20) fit that criteria, and superimposing those models to create a composite prediction. Peterson and Holt (2003) demonstrate how GARP models can be combined using GIS to examine variation in niche characteristics across the range of a species.

Guisan and Zimmerman (2000) characterized many of the modeling approaches currently in use as static and probabilistic, meaning that the models relate distributions of species to current environmental conditions without regard to future environmental change. Typically, only one data set is available to construct the model, so that statistical techniques such as jack-knifing, cross-validation, or boot-strapping are used to evaluate it. Ideally, one would like to have two independent data sets—one to build the model and one to evaluate it (Guisan and Zimmermann 2000).

Suggestions made more than 30 years ago that modeling approaches should be tested using data sets of known structure (Alldredge and Ratti 1986) have largely been ignored (O'Connor 2002). Most often, different models are compared using a single data set (e.g. Loiselle et al. 2003); only occasionally are models tested with two or more independent data sets (e.g. Mitchell et al. 2001, O'Connor and

Wagner 2004). Mitchell et al. (2001) found that microhabitat and landscape models worked equally as well in a managed forest in South Carolina, and combining the two models did little to improve the results. Models were refined by using other information on species distributions and analyzing a second independent data set. Models for Neotropical and short-distance migrants were better than models for permanent resident species, and models performed better for habitat specialists than for habitat generalists. In her study of birds associated with cottonwood (*Populus angustifolia*) riparian forests, Saab (1999) also concluded that landscape variables were better predictors than microhabitat (variation within cottonwood stands) and macrohabitat (variation between cottonwood stands) variables.

Use of Species Distribution Models in Conservation

There are two main applications of SDMs to conservation problems: forecasting future ranges of species in relation to environmental change and gap analyses. We first discuss each in turn, and then conclude this section by discussing several issues that apply specifically to the usefulness of these models with birds.

A common use of SDMs has been to evaluate the potential effects of future climate change. Species distribution models are used to create a climate envelope that relates recent species locations to current climates (e.g. temperature, precipitation, and seasonality) to estimate current distributional areas where populations persist in the face of competition, predation, and other challenges. Future distributions are then derived by applying the climate envelop to scenarios for future climate under the assumptions of different dispersal abilities. Such work has been conducted with Mexican birds, with birds of the U.S. Rocky Mountains and Great Plains, and with endemic Australian vertebrates (Peterson et al. 2002a, Peterson 2003a, Williams et al. 2003). Thomas et al. (2004) recently used this approach to explore the effects of different climate warming scenarios on future ranges and extinction for a variety of taxa. The ability of these models to accurately predict distributions 50 or more years into the future remains untested, and the validity of their predictions is unknown.

A variant of SDM has been used to predict the behavior of invasive species and the locations of future invasions. Using an approach called "climate-matching" (Peterson 2003b), species are assumed to be able to invade and establish populations only in areas that match the "niche" or ecological conditions where they occur in their native range. Species distribution models are first constructed and tested in the native range of the species, and then applied to other regions to forecast locations favorable for invasions (Peterson and Vieglais 2001, Peterson et al. 2002b). Peterson (2003b) provides an excellent review of this emerging approach.

Gap analyses are being conducted in each state as part of the Gap Analysis Program of the U.S. Geological Survey Gap Analysis Program (Jennings 2000). Regional analyses that merge multiple states are underway for several western states whose base maps are complete. This process will eventually produce a seamless map of biodiversity for most portions of the United States. The true value and power of gap analysis will then be realized, as ecoregions replace states as the units of analysis (e.g. Dietz and Czech 2005).

The source of available animal data typically dictates the scale that is chosen for gap-analysis maps. For example, the Arkansas Gap Analysis project (Smith et al. 1998) used counties as the smallest geographic area to map distributions of vertebrates (Fig. 5). Arkansas has 75 counties of about equal size, and the Arkansas Audubon Society has collected bird distribution data for portions of each county for >40 years. On the basis of county records and suitable habitat, the Ovenbird (*Seiurus aurocapilla*) is predicted to breed in the hardwoods of the Interior Highlands and on Crowley's Ridge in northeastern Arkansas, but not in the southern Gulf Coastal Plain, which has suitable hardwoods but is outside the known breeding range of the species in the state (Fig. 5). Eliminating habitats that the species never uses makes the predicted distribution more detailed than the original county-occurrence data. This technique can be extrapolated to the entire range of a species, providing a useful tool for refining range maps of species on the basis of use or nonuse of habitat types.

The strengths and weaknesses of species distribution models have recently been reviewed thoroughly by Scott et al. (2002), especially in the chapters by O'Connor (2002) and Van Horn

(2002). Here, we discuss two aspects that apply specifically to avian models and their usefulness in conservation of birds—scale and alternative analytical approaches.

Two aspects of scale need to be addressed in any SDM: (1) the scale at which various data sets were collected and analyzed (Holland et al. 2004) and (2) the extent to which data collected at a smaller scale are extrapolated to some larger scale (Miller et al. 2004). Selecting the wrong scale can lead to misleading or erroneous conclusions concerning how bird species are associated with various habitat characteristics (Mitchell et al. 2001). Birds, particularly migratory species, present a real challenge, because they may perceive the landscape on a scale from less than a hectare to thousands of square kilometers (Wiens et al. 2002). It is unlikely that one scale would be appropriate for assessing landscape patterns for all bird species within a region or even within groups of birds with similar ecologies (Mitchell et al. 2001). Conversely, Savignac et al. (2000) concluded that Pileated Woodpeckers (*Dryocopus pileatus*) in Quebec simultaneously required habitat features at two different scales.

The patch of habitat used in any avian landscape study is related to scale. Important aspects of patch dynamics are patch quality; characteristics of patch boundaries, particularly as they relate to species movements between adjacent patches; landscape connectivity, or how easily organisms can move across the landscape; and patch context, the characteristics of landscape (or adjacent patches) surrounding a particular patch (Wiens et al. 2002, Holland et al. 2004). For example, Kilgo et al. (1998) found differences in bird community structure in bottomland hardwoods in South Carolina depending on whether the forest patch was enclosed by pine (*Pinus taeda, P. palustris*) or by field-scrub habitats.

One potential problem in gap analysis that relates to patch size and scale is defining a realistic minimum mapping unit (MMU). Although the vegetation maps are typically derived from satellite imagery at a resolution of 30 m, the analysis of animals to the vegetation map is done at some larger scale (e.g. MMU of 40 ha or 100 ha). Classifying a patch of 100 ha or greater as a single vegetation type obscures all the vegetation heterogeneity within that patch. Choice of patch size potentially changes the results of a gap analysis (Karl et al. 2000). Another criticism of the gap-analysis approach is that models only link species distributions to patches at a specific time in a specific region (O'Connor 2002). Gap models, therefore, become outdated rather quickly and should be updated on some regular basis (e.g. redone every 10–20 years). A third criticism is that using vegetation as a surrogate for predicted distributions decreases resolution so that additional information needs to be gathered through more sampling of the species distributions (Stockwell and Peterson 2003).

O'Connor (2002) has suggested that a totally new approach is needed for the modeling of animal–habitat relationships, different from the current emphasis on correlations (see also Van Horn 2002). No doubt, the next decade will see the development of many other techniques as instrumentation and computer analyses increase in sophistication. Guisan and Zimmerman (2000) present a review of many new techniques for development of predictive species–habitat models that are an improvement over existing static models, and Ferrier et al. (2004) present a new global biodiversity modeling approach based on richness, faunal turnover, and species–area relationships. Here, we highlight three of those new approaches.

Classification and regression trees (CART) are an alternative to many statistical techniques now commonly used to model species relationships. As discussed in De'ath and Fabricus (2000) and O'Connor and Wagner (2004), trees explain variation of a single response variable by repeatedly splitting data into more homogeneous groups, using combinations of explanatory variables that can be either categorical or numeric. Each group is characterized by a typical value of the response variable, the number of observations in the group, and the values of the exploratory variables that define it. Results are presented graphically, which aids in understanding the relationships. De'ath and Fabricus (2000) list the advantages of this approach: (1) different types of response variables can be used; (2) it can be used for interactive exploration, description, and prediction; (3) it is invariant to transformations of explanatory variables; (4) the graphical results allow easy interpretation of complex ecological data, including interactions; (5) models can be selected by cross-validation; and (6) missing data values are not a problem.

Classification and regression trees have many applications in avian studies (reviewed in O'Connor and Wagner 2004). For example, O'Connor and Jones (1997) used CART to examine long-term trends in species occurrence on Breeding Bird Surveys (BBS) in the conterminous United States. Their analysis suggested that 15% of the BBS had lost an average of 17 species. Using CART analysis on Breeding Bird Census data from the United States, O'Connor and Wagner (2004) were able to generally confirm the results of O'Connor and Jones (1997). Becker and Beissinger (2003) found CART to be an excellent way to analyze the marine habitat characteristics that affected the at-sea distribution of Marbled Murrelets.

Use of generalized linear models (GLMs) and generalized additive models (GAMs) was the focus of a recent symposium (Guisan et al. 2002). Generalized linear models are mathematical extensions of linear models that do not force data into unnatural scales, allowing for nonlinearity and nonconstant variance structures in the data. Data can be assumed to be from several families of probability distributions, including normal, binomial, Poisson, and negative binomial. Generalized additive models have the underlying assumption that functions are additive and that the components are smooth. The advantage of GAMs is the ability to deal with highly nonlinear and nonmonotonic relationships between the response and the set of explanatory variables. Guisan et al. (2002) present a brief overview of the mathematical relations among linear regression, GLMs, and GAMs. Problems that may occur with these approaches include multicollinearity, limitations in using stepwise regression, and spatial autocorrelation (O'Connor 2002). Also, both CART and GAM models tend to be "data-hungry," precluding their use on relatively small data sets. Guisan et al. (2002) suggest that CART models may prove useful in identifying interactions among predictors used in GLMs and GAMs.

Another emerging approach is focal patch analysis (Brennan et al. 2002, Holland et al. 2004), which takes into account not only the patch in which a species occurs, but also the characteristics of the landscape surrounding the patch (i.e. patch context; Saab 1999). Data are collected on species abundance or richness in a number of patches or sites, and landscape predictor variables are measured in areas centered on the patch or site locations. Each patch then becomes a single data point in the analysis, and the influence of habitat variables measured at a large scale can be used to examine species abundance or richness. See Holland et al. (2004) for an example of this approach and the Landscape Ecology Laboratory website (see Acknowledgments) for downloadable programs to conduct this analysis. Important issues in this approach are (1) determining the appropriate landscape scale, (2) using multiple landscapes at multiple scales, (3) patch- and landscape-scale factors, and (4) tradeoffs of intensive sampling versus obtaining adequate sample sizes (see Brennan et al. 2002). Although it has not been used in any avian studies (J. Holland pers. comm.), Holland et al. (2004) estimate that many existing data sets could be re-analyzed using focal patch analysis.

What species distribution models lack in precision, they make up in generality. Products of SDMs can be considered working hypotheses and should be treated as such. The models are, in essence, a structured way to visualize those hypotheses, and should be viewed as starting points for discussion of the current and future spatial distribution of terrestrial vertebrates. The temporal dynamics of the landscape, coupled with the dynamics of real populations, may limit the usefulness of SDMs as a conservation planning tool for specific landscapes (Conroy and Noon 1996, Scott et al. 2002). They can be more useful in regional planning efforts (e.g. White et al. 1997), but should not be substituted for site-specific studies and field work (Scott et al. 1993, 2002; Peterson 2005).

Intelligent Use of Models to Make Conservation Decisions

Attributes of Useful Models

To aid conservation decisions, intelligent use of the models discussed here requires an understanding of their unique attributes. Fortunately, some generalities emerge that can be used to evaluate these models, as well as other model types that we have not discussed. Certain attributes will make models, regardless of type, more useful for making conservation decisions. These attributes are not easily incorporated into existing models, but they represent goals that modelers should strive to reach, so that field biologists will be encouraged to make greater use of, and provide data for, their models.

First, there is a necessary tradeoff between the generality and the realism of models (Levins 1969). Models that are general in structure lead to useful "rules of thumb" but often lack the realism needed for application to specific (conservation) issues. The compromise is between simple models with minimal data requirements but limited specificity and complex models with extensive data requirement but high specificity. The challenge is in building simple, feasible, and understandable models that nevertheless are detailed enough to address real problems.

In conservation, there is often the perception that increased realism is needed to apply models to particular situations. This can result in the construction of highly complex models that may be unreliable because of uncertainties in parameter values (e.g. dispersal behavior and unknown temporal and spatial variation in vital rates). A reasonable approach is to develop models of intermediate complexity that have enough realism to be applied to particular conservation problems but avoid unnecessary complexity—that is, apply the principle of parsimony (Burnham and Anderson 1992, 2002).

In the case of the Northern Spotted Owl, for example, an IBM was developed to evaluate competing plans for management of public lands in the Pacific Northwest of the United States. Because each plan could be expressed as a map showing the state of the landscape 100 years in the future, a spatially explicit model was used to rank the two alternatives. This model was quite complex, including age-structure, spatial subdivision, and individual variation in vital rates (McKelvey et al. 1993; Noon and McKelvey 1996a, b). Other researchers have suggested that simpler models excluding age-structure and spatial information (Wennergren et al. 1995) or using spatially explicit optimization procedures (Hof and Raphael 1997) would have provided similar insights.

Second, models should be mechanistic. Those used to support conservation decisions should be based on current understanding of the key processes contributing to a system's dynamics. Mechanistic models allow decision-makers to see how those processes are influenced by management options. The ability to provide a causal explanation will increase public understanding and make it easier to gain public support for costly conservation decisions. In addition, mechanistic models may provide useful predictions for application to situations outside the range of conditions on which the model was based.

Third, models should be forward-looking rather than predictive. The limited ability of models to accurately predict future events must be clearly stated. What actually occurs in the future is, at best, one of the many possible simulated trajectories of stochastic model output. As the time horizon of a model becomes longer, the uncertainty in its predictions becomes greater. Outputs from these models are better treated as heuristic expectations that presumably bound the range of possible outcomes.

Fourth, models should contain relevant variables, and their assumptions and results must be biologically realistic. Variables used as model input and the form of model output must be understandable and relevant to the conservation action. Presenting model results in technical and complex terms is certain to diminish their attractiveness to field biologists and their value to decision-makers. Unnecessary complexity increases the likelihood that data needs and model insights will be ignored.

Fifth, models should be testable. Their hypotheses about how a system operates must make predictions that can be tested by experiment or observation over both the short and the long term. For conservation purposes, this requires model output in the form of variables whose values can be measured in the field. Given a model-based projection of population response to management action, is that response observed? If not, the model and the decision must be revisited and revised accordingly.

Sixth, models should fully explore system uncertainty (Possingham et al. 2001, Burgman et al. 2005). For the foreseeable future, conservation decisions will be made with an incomplete understanding of system dynamics. It is the responsibility of the modeler to fully explore the model's uncertainty and make this known to decision-makers and field biologists. In the case of population models, the risks of specific management decisions must be made explicit to the public.

Seventh, modeling efforts benefit from constructing multiple models, because often there is limited agreement among researchers and decision-makers on how and at what rate nature works. Developing and testing competing models offers the opportunity to examine different mechanistic explanations of how populations respond to external factors. Ideally, the models should be structured differently so that they have the potential to produce qualitatively different projections. Each model's validity is evaluated relative to the existing data, and the one that is most aligned with existing data gains in credibility. Discriminating among competing models in terms of their relative concordance between prediction and observation is the essence of adaptive management.

It is difficult to develop models that have all these properties. For example, the more complex types of population models, SEPMs, often meet the last six of the above criteria, but fail on the first criterion of generality. A key point is to make models no more complex than they need to be for the problem under consideration or than the existing data allow, while retaining the mechanistic elements that make them useful in conservation. Finally, it is useful to consider the results of even the most mechanistic models not as precise predictions for the future but instead as heuristic expectations that provide a guide to the effects of management options.

Working with Multiple Models in a Decision-theoretic Framework

So far, the discussion of model types has not explicitly addressed how models can aid decision-making. Often, there is no obviously "correct" model of a population's or system's dynamics, and decisions about an appropriate model are often made in the context of incomplete information. In fact, multiple plausible models about how a system works may be in competition, each predicting a range of possible outcomes with different likelihoods. There may, however, be a "best" model in the sense of being most consistent with the available data. When competing scientific models are compared in the context of observable quantities or data, we move into the realm of statistical modeling (Nichols 2001). Decision theory, in the context of statistical model comparison, provides a step-down process to aid decision-making when there is uncertainty about the "true" structure and dynamics of the system being modeled (Possingham et al. 2001, Burgman 2005).

Several publications outline a process for making decisions about the management of dynamic systems that are poorly understood or for which there is little agreement on the factors that affect the state of the system (Hilborn and Mangel 1997, Burnham and Anderson 2002, Williams et al. 2002). The process, applicable to the conservation of imperiled bird species, for example, makes full use of available information and current understanding to select a single "best model" or a set of models with similar support from a data set (Burnham and Anderson 2002, Johnson and Omland 2004). The decision-making process requires the development of two or more competing models that bound the breadth of uncertainty and describe and project the dynamics of the population. Competing models share a common objective: supplying the information needed for making the management decision that maximizes the likelihood of realizing a goal within some set of constraints. Data relevant to the state of the system (e.g. population size and reproductive rates) are collected, and a measure of the probability of the data, given that the model is true, is computed. Through application of Bayes's Theorem (Hilborn and Mangel 1997, Johnson and Omland 2004), this measure is then turned on its head and interpreted as a measure of the chance that the model is the appropriate description of the system, given the data. To determine which is the best descriptor of the population's dynamics, models are compared in terms of their posterior likelihoods, given the data. That is, quantitative predictions of the models are compared with observations provided by an ongoing monitoring program. Insights from the current best model are then used to inform the conservation decision process.

The process of model comparison is iterative, depending on a regular assessment of the state of the population. After each period of data collection, model predictions are again compared, and the current state of the system and the posterior probabilities of being the correct model are recomputed for each model. The current best model is then used as the basis for decision-making until the next assessment occurs. Management of hunted waterfowl provides an excellent example of the use of decision-theoretic models for conservation. The U.S. Fish and Wildlife Service uses the competing model procedure to make management decisions regarding optimal harvest of waterfowl populations (Nichols et al. 1995, Johnson et al. 1997, Johnson and Williams 1999). In this case, the competing models bracket the range of uncertainty associated with random environmental variation, incomplete control over harvest, and uncertainty about the biological mechanisms that control population size.

A full discussion of the decision-theoretic approach and its uses in the adaptive management process is beyond the scope of this review. However, it is apparent that these methods will be more commonly used in the future (Clark 2005). Hilborn and Mangel (1997), Burnham and Anderson (2002), and particularly Williams et al. (2002) provide a good introduction to these methods.

How Field Biologists Can Interact with Modelers to Improve Conservation Decisions

Ecology has been described as a science of case studies and rough generalizations rather than of general theory and exceptionless empirical laws (Shrader-Frechette and McCoy 1993). Despite the heuristic power of an ecological perspective, there is little general theory to provide policy-related predictions. Given the imprecision of ecological theory, it offers limited guidance to adjudicate public conflicts over conservation decisions. Conflict will remain part of conservation practice as long as divisions continue in society over the importance of biological diversity, our ethical responsibility to future human generations, and the severity of the conservation crisis. In an arena where consensus is difficult, any tools that allow us to project the possible consequences of present decisions should be fully exploited. The diversity and complexity of models employed in avian conservation is increasing rapidly, and their application is becoming more widespread. Use of models in avian conservation may not be a panacea, but neither is it a passing fancy. Thus, modelers and field biologists are becoming increasingly dependent on one another to achieve effective conservation.

The evolution of models during the era of the personal computer—from limited ability to incorporate complex biological details to spatial and mechanistic models that assume detailed knowledge of an animal's behavior—promotes synergy between modelers and field biologists. Mechanistic models that incorporate more biological details may be more appealing to field biologists, but they increase the burden on field biologists to collect more kinds of data and to verify that the biological processes incorporated in models are correct. Does survival vary with density in the manner depicted? Are the movement rules in the model correct? Are the measures and correlates used to estimate annual variation in fecundity sufficient? Indeed, some models are too appealing. Increasingly, modelers must caution field biologists about their reliance on complex models that incorporate great biological detail, whereas field biologists previously cautioned modelers about their reliance on models that were too simple biologically.

Collaboration between field biologists and modelers is often a necessity for effective conservation. Field biologists need modelers to depict what they know about a system in an appropriate way within a formal framework. Modelers need field biologists to provide data necessary for parameterization, to test whether processes incorporated in the model are correct, and to perform model validation. A model can guide collection of data and testing of mechanisms. Data can guide selection of the most appropriate model and determine the appropriate level of complexity (Burnham and Anderson 2002). Facilitating the link between modelers and field biologists are the statisticians. They play an essential role in designing methods of data collection and analyses needed to estimate parameters and compare model predictions with observed outcomes. Field biologists can remind modelers, statisticians, and policy makers of the weaknesses of particular models. Modelers and statisticians can indicate to field biologists the deficiencies in existing data. Productive collaborations of this sort involving one or more of the authors have promoted conservation efforts with Northern Spotted Owls, Red-cockaded Woodpeckers, Marbled Murrelets, Snail Kites, and Florida Scrub-Jays. Field biologists can make great contributions to conservation by participating in such collaborations.

A frequent theme of this monograph has been the onerous data requirements of new, complex types of models. The construction and application of population models for informing conservation decisions has been greatly hindered by the paucity of species that have received long-term field studies of population dynamics (Shaffer et al. 2002), and by the lack of standards and a repository for demographic databases. In the case of population models (e.g. spatially explicit models, metapopulation models, and stochastic single-population models), the information required consists largely of typical demographic and behavioral data collected by field biologists. However, the specificity of information required (e.g. age- or stage-specific demographic rates estimated with respect to landscape characteristics), the duration of time needed to produce accurate variance estimates for vital rates, and the difficulties inherent in accurately estimating survival rates make large sample sizes and long-term field studies a necessity. Furthermore, some types of data required by new population models have rarely been collected, notably information about dispersal and movement behavior. When incorporated into models, estimates of dispersal parameters are typically based on little or no data. Again, the burden is on field biologists to produce these data if the predictive ability of models is to be improved. Finally, use of increasingly more mechanistic models results in further demands on field biologists to provide new kinds of data and experimentally confirm postulated mechanisms. Particularly important is information about variation in vital rates, especially as it relates to annual variation, variation with habitat type, and variation with landscape features (e.g. proximity to edge). Field biologists need to keep potentially important sources of variation in mind when designing their sampling scheme and analyzing their data.

All models have key assumptions. Complex models may often make fewer simplifying assumptions than simple models, but they demand more data. The validity of basic simplifying assumptions is an important issue with species distribution models and deterministic matrix population models. Key assumptions about dispersal patterns and mortality associated with movement plague metapopulation and spatially explicit models. In these cases, field biologists can make major contributions by collecting data to test key model assumptions.

Validation is a major issue for most of the models discussed. Models can be validated by testing primary (e.g. population projections) or secondary (e.g. dispersal distributions) predictions, or by evaluating the ability of the model to replicate past system behavior (Bart 1995). Species distribution models can be validated from presence–absence data collected in field surveys. Validation of population models and genetic models is more problematic. Population projections resulting from stochastic single-population, metapopulation, and spatially explicit population models are difficult to validate directly, but one can test various secondary predictions of the models to evaluate their general performance. For example, an SEPM might predict the proportion of young produced on a territory that are recruited into the breeding population to be a function of territory isolation, a prediction that could readily be tested. If such a prediction proved false, it might indicate that a simpler (i.e. not spatially explicit) model is sufficient to capture the dynamics of the population. Applying the models to real landscapes to compare simulated and observed population dynamics may be especially instructive in this regard (e.g. Lindenmayer et al. 2003, Schiegg et al. 2005). Sensitivity analysis is also useful in directing attention to the parameters that matter most to model performance. For complex models, identifying important variables to improve model accuracy and to guide field studies is the most significant use of sensitivity analysis.

Models can be useful tools for field biologists, as well as managers, if employed wisely. Parameter estimates for complex models are typically too poor, and our understanding of nature too incomplete, to permit one to have much faith in specific predictions, such as the number of years it will take a population to become extinct, or the size of a population 50 or 100 years into the future. But models can be used to ask whether, given our current understanding of the system as depicted in the model, a population will persist longer, or be larger, under one management approach than under another (Beissinger and Westphal 1998, McCarthy et al. 2003, Lotts et al. 2004). Ideally, these assessments would then be tested through an adaptive management process (Walters 1986, Ludwig and Walters 2002). The best conservation decisions will occur where cooperative interaction enables field biologists, modelers, statisticians, and managers to contribute effectively.

Acknowledgments

We are greatly indebted to C. Thompson and V. Nolan, Jr., for encouragement and numerous editorial improvements of the work leading to this publication. S.R.B. acknowledges the National Science Foundation for grant support; the American Ornithologists' Union for providing the venue for the symposium that catalyzed this monograph; and G. White, B. Noon, and the Department of Fishery and Wildlife Biology at Colorado State University for hospitality and office space during the final revisions of this work. K.G.S. and D.G.C. thank the hundreds of people involved with the Arkansas Gap Analysis project (www.cast.uark.edu/gap). Their research was conducted at the Center for Advanced Spatial Technologies at the University of Arkansas, which is directed by W. F. Limp. R. Dzur was especially helpful with that project. Funding for this research and participation in the symposium was provided through the Arkansas Cooperative Fish and Wildlife Research Unit by U.S. Fish and Wildlife Service award number 14-16-0009-1567, W013 to K.G.S. and W. F. Limp. Part of this manuscript was written while K.G.S. was a Bullard Fellow at the Harvard Forest. MARK is available at www.warnercnr.colostate.edu/~gwhite/mark/mark.htm. PRESENCE is available at www.mbr-pwrc.usgs.gov/software/doc/presence/presence.html. The Landscape Ecology Laboratory website is at www.carleton.ca/lands-ecol/.

Literature Cited

Aber, J. D. 1997. Why don't we believe in models? Bulletin of the Ecological Society of America 78:232–233.

Akçakaya, H. R. 2000a. Conservation and management for multiple species: Integrating field research and modeling into management decisions. Environmental Management 26 (Supplement):75–S83.

Akçakaya, H. R. 2000b. Population viability analyses with demographically and spatially structured models. Ecological Bulletin 48:23–28.

Akçakaya, H. R. 2004. RAMAS METAPOP 5.0: Viability Analysis for Stage Structured Metapopulations. Applied Biomathematics, Setauket, New York.

Akçakaya, H. R., and J. L. Atwood. 1997. A habitat-based metapopulation model of the California Gnatcatcher. Conservation Biology 11:422–434.

Akçakaya, H. R., M. A. McCarthy, and J. L. Pearce. 1995. Linking landscape data with population viability analysis: Management options for the Helmeted Honeyeater *Lichenostomus melanops cassidix*. Biological Conservation 73:169–176.

Alldredge, J. R., and J. T. Ratti. 1986. Comparison of some statistical techniques for analysis of resource selection. Journal of Wildlife Management 50:157–165.

Anderson, D. R. 1975. Optimal exploitation strategies for an animal population in a Markovian environment: A theory and an example. Ecology 56:1281–1297.

Anderson, R. P., D. Lew, and A. T. Peterson. 2003. Evaluating predictive models of species' distributions: Criteria for selecting optimal models. Ecological Modelling 162:211–232.

Andren, J. 1994. Effects of habitat fragmentation on birds and mammals in landscapes with different proportions of suitable habitat: A review. Oikos 70:355–366.

Ariño, A., and S. L. Pimm. 1995. On the nature of population extremes. Evolutionary Ecology 9: 429–443.

Avise, J. C., and J. L. Hamrick. 1996. Conservation Genetics: Case Histories from Nature. Columbia University Press, New York.

Ballou, J. D., and R. C. Lacy. 1995. Identifying genetically important individuals for management of genetic variation in pedigreed populations. Pages 76–111 *in* Population Management for Survival and Recovery: Analytical Methods and Strategies in Small Population Conservation (J. D. Ballou, M. Gilpin, and T. J. Foose, Eds.). Columbia University Press, New York.

Ballou, J. D., and K. Ralls. 2004. Genetic status and management of California Condors. Condor 106:215–228.

Bart, J. 1995. Amount of suitable habitat and viability of Northern Spotted Owls. Conservation Biology 9:943–946.

Becker, B. H., and S. R. Beissinger. 2003. Scale-dependent habitat selection by a nearshore seabird, the Marbled Murrelet, in a highly dynamic upwelling system. Marine Ecology Progress Series 256:243–255.

Beissinger, S. R. 1986. Demography, environmental uncertainty, and the evolution of mate desertion in the Snail Kite. Ecology 67:1445–1459.

Beissinger, S. R. 1995a. Population trends of the Marbled Murrelet projected from demographic analyses. Pages 385–393 *in* Ecology and Conservation of the Marbled Murrelet (C. J. Ralph, G. L. Hunt, Jr., M. G. Raphael, and J. F. Piatt, Eds.). U.S. Department of Agriculture, Forest Service General Technical Report PSW-GTR 152.

Beissinger, S. R. 1995b. Modeling extinction in periodic environments: Everglades water levels and Snail Kite population viability. Ecological Applications 5:618–631.

Beissinger, S. R. 2000. Ecological mechanisms of extinction. Proceedings of the National Academy of Sciences USA 97:11688–11689.

Beissinger, S. R., and J. P. Gibbs. 1993. Are variable environments stochastic? A review of methods to quantify environmental predictability. Pages 132–146 *in* Adaptation in Stochastic

Environments (J. Yoshimura, and C. W. Clark, Eds.). Springer-Verlag, Berlin.

Beissinger, S. R., and D. R. McCullough. 2002. Population Viability Analysis. University of Chicago Press, Chicago.

Beissinger, S. R., and N. Nur. 1997. Appendix B: Population trends of the Marbled Murrelet projected from demographic analysis. Pages B1–B35 *in* Recovery Plan for the Threatened Marbled Murrelet (*Brachyramphus marmoratus*) in Washington, Oregon, and California. U.S. Fish and Wildlife Service, Portland, Oregon.

Beissinger, S. R., and M. I. Westphal. 1998. On the use of demographic models of population viability in endangered species management. Journal of Wildlife Management 62:821–841.

Belisle, M. 2005. Measuring landscape connectivity: The challenge of behavioral landscape ecology. Ecology 86:1988–1995.

Belovsky, G. E., C. Mellison, C. Larson, and P. A. Van Zandt. 2000. Experimental studies of extinction dynamics. Science 286:1175–1177.

Blouin, M. S. 2003. DNA-based methods for pedigree reconstruction and kinship analysis in natural populations. Trends in Ecology and Evolution 18:503–511.

Bolker, B. M. 2004. Continuous-space models for population dynamics. Pages 45–72 *in* Ecology, Genetics, and Evolution of Metapopulations (I. Hanski, and O. Gaggiotti, Eds.). Elsevier Academic Press, New York.

Boone, R. B., and M. L. Hunter. 1996. Using diffusion models to simulate the effects of land use on grizzly bear dispersal in the Rocky Mountains. Landscape Ecology 11:51–64.

Boone, R. B., and W. B. Krohn. 1999. Modeling the occurrence of bird species: Are the errors predictable? Ecological Applications 9:835–848.

Bradstock, R. A., M. Bedward, J. Scott, and D. A. Keith. 1996. Simulation of the effect of spatial and temporal variation in fire regimes on the population viability of a *Banksia* species. Conservation Biology 10:776–784.

Brennan, J. M., D. J. Bender, T. A. Contreras, and L. Fahing. 2002. Integrating landscape ecology into natural resource management. Pages 68–91 *in* Focal Patch Landscape Studies in Wildlife Management: Optimizing Sampling Effort across Scales (J. Liu and W. W. Taylor, Eds.). Cambridge University Press, New York.

Brook, B. W., M. A. Burgman, and R. Frankham. 2000a. Differences and congruencies between PVA packages: The importance of sex ratio for predictions of extinction risk. [Online.] Conservation Ecology 4:6. Available at www.ecologyandsociety.org/vol4/iss1.

Brook, B. W., and J. Kikkawa. 1998. Examining threats faced by island birds: A population viability analysis on the Capricorn Silvereye using long-term data. Journal of Applied Ecology 35:491–503.

Brook, B. W., J. J. O'Grady, A. P. Chapman, M. A. Burgman, H. R. Akçakaya, and R. Frankham. 2000b. Predictive accuracy of population viability analysis in conservation biology. Nature 404:385–387.

Brooks, T. M., M. I. Bakarr, T. Boucher, G. A. B. da Fonseca, C. Hilton-Taylor, J. M. Hoekstra, T. Moritz, S. Olivieri, J. Parrish, R. L. Pressy, and others. 2004. Coverage provided by the global protected-area system: Is it enough? BioScience 54:1081–1091.

Bruna, E. M., and M. K. Oli. 2005. Demographic effects of habitat fragmentation on a tropical herb: Life-table response experiments. Ecology 86:1816–1824.

Brys, R., H. Jacquemyn, P. Endels, G. De Blust, and M. Hermy. 2005. Effect of habitat deterioration on population dynamics and extinction risks in a previously common perennial. Conservation Biology 19:1633–1643.

Burgman, M. A. 2005. Risks and Decisions for Conservation and Environmental Management. Cambridge University Press, Cambridge, United Kingdom.

Burgman, M. A., S. Ferson, and H. R. Akçakaya. 1993. Risk Assessment in Conservation Biology. Chapman and Hall, London.

Burgman, M. A., D. B. Lindenmayer, and J. Elith. 2005. Managing landscapes for conservation under uncertainty. Ecology 86:2007–2017.

Burnham, K. P., and D. R. Anderson. 1992. Data-based selection of an appropriate biological model: The key to modern data analysis. Pages 16–30 *in* Wildlife 2001: Populations (D. R. McCullough, and R. H. Barrett, Eds.). Elsevier Applied Science, New York.

Burnham, K. P., and D. R. Anderson. 2002. Model Selection and Multimodel Inference: A Practical Information-theoretic Approach, 2nd ed. Springer-Verlag, New York.

Busby, J. R. 1991. BIOCLIM—A bioclimatic analysis and prediction system. Pages 64–68 *in* Nature Conservation: Cost Effective Biological Surveys and Data Analysis (C. R. Margules and M. P. Austin, Eds.). CSIRO, Canberra, Australia.

Bustamante, J. 1996. Population viability analysis of captive and released Bearded Vulture populations. Conservation Biology 10:822–831.

Butterfield, B. R., B. Csuti, and J. M. Scott. 1994. Modeling vertebrate distributions for gap analysis. Pages 53–68 *in* Mapping the Diversity of Nature (R. I. Miller, Ed.). Chapman and Hall, London.

Cam, E., L. Lougheed, R. Bradley, and F. Cooke. 2003. Demographic assessment of a Marbled

Murrelet population from capture–recapture data. Conservation Biology 17:1118–1126.

Carpenter, G., A. N. Gillison, and J. Winter. 1993. DOMAIN—A flexible modelling procedure for mapping potential distributions of plants and animals. Biodiversity Conservation 2:667–680.

Caswell, H. 1996a. Second derivatives of population growth rate: Calculation and applications. Ecology 77:870–879.

Caswell, H. 1996b. Analysis of life table response experiments. II. Alternative parameterizations for size- and stage-structured models. Ecological Modelling 88:73–82.

Caswell, H. 2000. Prospective and retrospective perturbation analyses: Their roles in conservation biology. Ecology 81:619–627.

Caswell, H. 2001. Matrix Population Models: Construction, Analysis, and Interpretation, 2nd ed. Sinauer Associates, Sunderland, Massachusetts.

Caughley, G., and A. Gunn. 1996. Conservation Biology in Theory and Practice. Blackwell Science, Cambridge, Massachusetts.

Charlesworth, D., and B. Charlesworth. 1987. Inbreeding depression and its evolutionary consequences. Annual Review of Ecology and Systematics 18:237–268.

Charnov, E. L. 1993. Life History Invariants: Some Explorations of Symmetry in Evolutionary Ecology. Oxford University Press, Oxford.

Citta, J. J., and L. S. Mills. 1999. What do demographic sensitivity analyses tell us about controlling Brown-headed Cowbirds. Pages 121–134 *in* Research and Management of the Brown-headed Cowbird in Western Landscapes (M. L. Morrison, L. S. Hall, S. K. Robinson, S. I. Rothstein, D. Caldwell Hahn, and T. D. Rich, Eds.). Studies in Avian Biology, no. 18.

Clark, J. S. 2005. Why environmental scientists are becoming Bayesians. Ecology Letters 8:2–14.

Colwell, R. K. 1974. Predictability, constancy and contingency of periodic phenomena. Ecology 55:1148–1153.

Conroy, M. J., Y. Cohen, F. C. James, Y. G. Matsinos, and B. A. Maurer. 1995. Parameter estimation, reliability, and model improvement for spatially explicit models of animal populations. Ecological Applications 5:17–19.

Conroy, M. J., and B. R. Noon. 1996. Mapping of species richness for conservation of biological diversity: Conceptual and methodological issues. Ecological Applications 6:763–773.

Cooch, E., R. F. Rockwell, and S. Brault. 2001. Retrospective analysis of demographic responses to environmental change: A Lesser Snow Goose example. Ecological Monographs 71:377–400.

Cook, R. R., and I. Hanski. 1995. On expected lifetimes of small-bodied and large-bodied species of birds on islands. American Naturalist 145:307–315.

Cooper, C. B., J. R. Walters, and J. Priddy. 2002. Landscape patterns and dispersal success: Simulated population dynamics in the Brown Treecreeper. Ecological Applications 12: 1576–1587.

Corsi, F., J. de Leeuw, and A. Skidmore. 2000. Modeling species distributions with GIS. Pages 389–434 *in* Research Techniques in Animal Ecology: Controversies and Consequences (L. Boitani and T. K. Fuller, Eds.). Columbia University Press, New York.

Courchamp, F., T. Clutton-Brock, and B. Grenfell. 1999. Inverse density dependence and the Allee effect. Trends in Ecology and Evolution 14:405–410.

Cross, P. C., and S. R. Beissinger. 2001. Using logistic regression to analyze the sensitivity of PVA models: A comparison of methods based on African wild dog models. Conservation Biology 15:1335–1346.

Crouse, D. T., L. B. Crowder, and H. Caswell. 1987. A stage-based population model for loggerhead sea turtles and implications for conservation. Ecology 68:1412–1423.

Crowder, L. B., D. T. Crouse, S. S. Heppell, and T. H. Martin. 1994. Predicting the impact of turtle excluder devices on loggerhead sea turtle populations. Ecological Applications 4: 437–445.

Curnutt, J. L., J. Comiskey, M. P. Nott, and L. J. Gross. 2000. Landscape-based spatially explicit species index models for Everglades restoration. Ecological Applications 10:1849–1860.

da Fonseca, G. A. B., A. Balmford, C. Bibby, L. Boitani, F. Corsi, T. Brooks, C. Gascon, S. Olivieri, R. A. Mittermeier, N. Burgess, and others. 2000. It's time to work together and stop duplicating conservation efforts…following Africa's lead in setting priorities. Nature 405:393–394.

Dale, V. H., Ed. 2003. Ecological Modeling for Resource Management. Springer-Verlag, New York.

Daniels, S. J., and J. R. Walters. 2000. Inbreeding depression and natal dispersal in Red-cockaded Woodpeckers. Condor 102:482–491.

Davis, L. S., K. N. Johnson, P. S. Bettinger, and T. E. Howard. 2001. Forest Management: To Sustain Ecological, Economic, and Social Values, 4th ed. McGraw-Hill, Boston.

DeAngelis, D. L., and K. A. Rose. 1992. Which individual-based approach is most appropriate for a given problem? Pages 67–87 *in* Individual-Based Models and Approaches in Ecology: Populations, Communities, and

Ecosystems (D. L. DeAngelis and L. J. Gross, Eds.). Chapman and Hall, New York.

DeAngelis, D. L., K. A. Rose, L. B. Crowder, E. A. Marschall, and D. Lika. 1993. Fish cohort dynamics: Applications of complementary modeling approaches. American Naturalist 142:604–622.

De'ath, G., and K. E. Fabricus. 2000. Classification and regression trees: A powerful yet simple technique for ecological data analysis. Ecology 81:3178–3192.

de Kroon, H., A. Plaisier, J. van Groenendael, and H. Caswell. 1986. Elasticity: The relative contribution of demographic parameters to population growth rate. Ecology 67:1427–1431.

de Kroon, H., J. van Groenendael, and J. Ehrlén. 2000. Elasticities: A review of methods and model limitations. Ecology 81:607–618.

Dennis, B., P. L. Munholland, and J. M. Scott. 1991. Estimation of growth and extinction parameters for endangered species. Ecological Monographs 61:115–143.

Dietz, R. W., and B. Czech. 2005. Conservation deficits for the continental United States: An ecosystem gap analysis. Conservation Biology 19:1478–1487.

Doak, D., P. Kareiva, and B. Klepetka. 1994. Modeling population viability for the desert tortoise in the western Mojave Desert. Ecological Applications 4:446–460.

Doherty, P. F., E. A. Schreiber, J. D. Nichols, J. E. Hines, W. A. Link, G. A. Schenk, and R. W. Schreiber. 2004. Testing life history predictions in a long-lived seabird: A population matrix approach with improved parameter estimation. Oikos 105:606–618.

Donovan, T. M., R. H. Lamberson, A. Kimber, F. R. Thompson III, and J. Faaborg. 1995. Modeling the effects of habitat fragmentation on source and sink demography of Neotropical migrant birds. Conservation Biology 9:1396–1407.

Downer, R. 1993. GAPPS User Manual. Applied Biomathematics, Setauket, New York.

Drechsler, M., and M. A. Burgman. 2004. Combining population viability analysis with decision analysis. Biodiversity and Conservation 13:115–139.

Dreitz, V. J., J. D. Nicholas, J. E. Hines, R. E. Bennets, D. W. Kitchen, and D. L. DeAngelis. 2002. The use of resighting data to estimate the rate of population growth of the Snail Kite in Florida. Journal of Applied Statistics 29: 609–623.

Dunning, J. B., Jr., B. J. Danielson, and H. R. Pulliam. 1992. Ecological processes that affect populations in complex landscapes. Oikos 65: 169–175.

Dunning, J. B., Jr., B. J. Danielson, B. D. Watts, J. Liu, and D. G. Krementz. 2000. Studying wildlife at local and landscape scales: Bachman's Sparrows at the Savannah River Site. Pages 75–80 *in* Avian Research at the Savannah River Site: A Model for Integrating Basic Research and Long-term Management (J. B. Dunning, Jr., and J. C. Kilgo, Eds.). Studies in Avian Biology, no. 21.

Dunning, J. B., Jr., D. J. Stewart, B. J. Danielson, B. R. Noon, T. L. Root, R. H. Lamberson, and E. E. Stevens. 1995. Spatially explicit population models: Current forms and future uses. Ecological Applications 5:3–11.

Earnhardt, J. M., A. Lin, L. J. Faust, and S. D. Thompson. 2004. ZOORISK: A Risk Assessment Tool, version 1.0. Lincoln Park Zoo, Chicago.

Edwards, T. C., Jr., E. T. Deshler, D. Foster, and G. G. Moisen. 1996. Adequacy of wildlife habitat relation models for estimating spatial distributions of terrestrial vertebrates. Conservation Biology 10:263–270.

Emlen, J. M. 1989. Terrestrial population models for ecological risk assessment: A state-of-the-art review. Environmental Toxicology and Chemistry 8:831–842.

Engen, S., B. E. Sæther, and A. P. Møller. 2001. Stochastic population dynamics and time to extinction of a declining population of Barn Swallows. Journal of Animal Ecology 70: 789–797.

Etienne, R. S., C. J. F. ter Braak, and C. C. Vos. 2004. Application of stochastic patch occupancy models to real metapopulations. Pages 105–132 *in* Ecology, Genetics, and Evolution of Metapopulations (I. Hanski and O. Gaggiotti, Eds.). Elsevier Academic Press, New York.

Fahrig, L. 1992. Relative importance of spatial and temporal scales in a patchy environment. Theoretical Population Biology 41:300–314.

Fahrig, L. 1997. Relative effects of habitat loss and fragmentation on population extinction. Journal of Wildlife Management 61:603–610.

Fahrig, L., and G. Merriam. 1994. Conservation of fragmented populations. Conservation Biology 8:50–59.

Ferrier, S., G. V. N. Powell, K. S. Richardson, G. Manion, J. M. Overton, T. F. Allnutt, S. E. Cameron, K. Mantle, N. D. Burgess, D. P. Faith, and others. 2004. Mapping more of terrestrial biodiversity for global conservation assessment. BioScience 54:1101–1109.

Fieberg, J., and S. P. Ellner. 2000. When is it meaningful to estimate an extinction probability? Ecology 81:2040–2047.

Fielding, A. H., and J. F. Bell. 1997. A review of methods for the assessment of prediction errors in conservation presence/absence models. Environmental Conservation 24:38–49.

Fleishman, E., C. Ray, P. Sjogren-Gulve, C. L. Boggs, and D. D. Murphy. 2002. Assessing the roles of patch quality, area, and isolation in predicting metapopulation dynamics. Conservation Biology 16:706–715.

Fleming, D. M., W. F. Wolff, and D. L. DeAngelis. 1994. Importance of landscape heterogeneity to Wood Storks in Florida Everglades. Environmental Management 18:743–757.

Foley, P. 1994. Predicting extinction times from environmental stochasiticity and carrying capacity. Conservation Biology 8:124–137.

Frankham, R. 1995. Effective population size-adult population size ratios in wildlife: A review. Genetical Research 66:95–107.

Franklin, A. B., D. R. Anderson, R. J. Gutiérrez, and K. P. Burnham. 2000. Climate, habitat quality, and fitness in Northern Spotted Owl populations in Northwestern California. Ecological Monographs 70:539–590.

Franklin, A. B., R. J. Gutiérrez, J. D. Nichols, M. E. Seamans, G. C. White, G. S. Zimmerman, J. E. Hines, T. E. Munton, W. S. LaHaye, J. A. Blakesley, and others. 2004. Population dynamics of the California Spotted Owl (*Strix occidentalis occidentalis*): A meta-analysis. Ornithological Monographs, no. 54.

Franklin, I. R. 1980. Evolutionary change in small populations. Pages 135–150 *in* Conservation Biology: An Evolutionary–Ecological Perspective (M. E. Soulé and B. A. Wilcox, Eds.). Sinauer Associates, Sunderland, Massachusetts.

Franklin, I. R., and R. Frankham. 1998. How large must populations be to retain evolutionary potential? Animal Conservation 1:69–73.

Gautschi, B., G. Jacob, J. J. Negro, J. A. Godoy, J. P. Muller, and B. Schmid. 2003. Analysis of relatedness and determination of the source of founders in the captive Bearded Vulture, *Gypaetus barbatus*, population. Conservation Genetics 4:479–490.

Geis, A. D., R. K. Martinson, and D. R. Anderson. 1969. Establishing hunting regulations and allowable harvest of Mallards in the United States. Journal of Wildife Management 33: 848–859.

Gerber, L. R., and G. R. VanBlaricom. 2001. Implications of three viability models, for the conservation status of the western population of Steller sea lions (*Eumetopias jubatus*). Biological Conservation 102:261–269.

Gertseva, V. I., and V. V. Gertseva. 2004. Classification of mathematical models in ecology. Ecological Modelling 178:329–334.

Getz, W. M., and R. G. Haight. 1989. Population Harvesting: Demographic Models of Fish, Forest, and Animal Resources. Princeton University Press, Princeton, New Jersey.

Gilligan, D. M., L. M. Woodworth, M. E. Montgomery, D. A. Briscoe, and R. Frankham. 1997. Is mutation accumulation a threat to the survival of endangered populations? Conservation Biology 11:1235–1241.

Gilpin, M. 1996. Metapopulations and wildlife conservation: Approaches to modeling spatial structure. Pages 11–27 *in* Metapopulations and Wildlife Conservation (D. R. McCullough, Ed.). Island Press, Washington, D.C.

Ginzburg, L. R., S. Ferson, and H. R. Akçakaya. 1990. Reconstructibility of density dependence and the conservative assesment of extinction risks. Conservation Biology 4:63–70.

Ginzburg, L. R., L. B. Slobodkin, K. Johnson, and A. G. Bindman. 1982. Quasiextinction probabilities as a measure of impact on population growth. Risk Analysis 2:171–181.

Godown, M. E., and A. T. Peterson. 2000. Preliminary distributional analysis of U.S. endangered bird species. Biodiversity Conservation 9:1313–1322.

Goodman, D. 1987. The demography of chance extinction. Pages 11–34 *in* Viable Populations for Conservation (M. E. Soulé, Ed.). Cambridge University Press, Cambridge, United Kingdom.

Goodman, D. 2002. Predictive Bayesian population analysis: A logic for listing criteria, delisting criteria and recovery plans. Pages 447–469 *in* Population Viability Analysis (S. R. Beissinger and D. R. McCullough, Eds.). University of Chicago Press, Chicago.

Gould, W. R., and J. D. Nichols. 1998. Estimation of temporal variability of survival in animal populations. Ecology 79:2531–2538.

Grand, J., and S. R. Beissinger. 1997. When relocation of loggerhead sea turtle (*Caretta caretta*) nests becomes a useful strategy. Journal of Herpetology 31:428–434.

Grant, A., and T. G. Benton. 2000. Elasticity analysis for density-dependent populations in stochastic environments. Ecology 81:680–693.

Grant, P. R. 1986. Ecology and Evolution of Darwin's Finches. Princeton University Press, Princeton, New Jersey.

Green, R. E., and G. J. M. Hirons. 1991. The relevance of population studies to conservation of threatened birds. Pages 594–636 *in* Bird Population Studies: Relevance to Conservation and Management (C. M. Perrins, J.-D. Lebreton, and G. J. M. Hirons, Eds.). Oxford University Press, Oxford.

Griffith, A. B., and I. N. Forseth. 2005. Population matrix models of *Aeschynomene virginica*, a rare annual plant: Implications for conservation. Ecological Applications 15:222–233.

Grimm, V., H. Lorek, J. Finke, F. Koester, M. Malachinski, M. Sonnenschein, A. Moilanen,

I. Storch, A. Singer, C. Wissel, and K. Frank. 2004. META-X: Generic software for metapopulation viability analysis. Biodiversity and Conservation 13:165–188.

Grimm, V., and C. Wissel. 2004. The intrinsic mean time to extinction: A unifying approach to analysing persistence and viability of populations. Oikos 105:501–511.

Groom, M. J., and M. A. Pascual. 1998. The analysis of population persistence: An outlook on the practice of viability analysis. Pages 4–27 *in* Conservation Biology (P. L. Fiedler and P. M. Kareiva, Eds.). Chapman and Hall, New York.

Guisan, A., T. C. Edwards, Jr., and T. Hastie. 2002. Generalized linear and generalized additive models in studies of species distributions: Setting the scene. Ecological Modelling 157: 89–100.

Guisan, A., and N. E. Zimmermann. 2000. Predictive habitat distribution models in ecology. Ecological Modelling 135:147–186.

Haig, S. M. 1998. Molecular contributions to conservation. Ecology 79:413–425.

Haig, S. M., and J. D. Ballou. 2002. Pedigree models for managing small populations in the wild. Pages 388–405 *in* Population Viability Analysis (S. R. Beissinger, and D. R. McCullough, Eds.). University of Chicago Press, Chicago.

Haig, S. M., J. D. Ballou, and N. J. Casna. 1994a. Identification of kin structure among Guam Rail founders: A comparison of pedigrees and DNA profiles. Molecular Ecology 3:109–119.

Haig, S. M., J. D. Ballou, and N. J. Casna. 1995. Genetic identification of kin in Micronesian Kingfishers. Journal of Heredity 86:423–431.

Haig, S. M., J. D. Ballou, and S. R. Derrickson. 1990. Management options for preserving genetic diversity: Reintroduction of Guam Rails to the wild. Conservation Biology 4: 290–300.

Haig, S. M., J. R. Belthoff, and D. H. Allen. 1993a. Examination of population structure in Red-cockaded Woodpeckers using DNA profiles. Evolution 47:185–194.

Haig, S. M., J. R. Belthoff, and D. H. Allen. 1993b. Population viability analysis for a small population of Red-cockaded Woodpeckers and an evaluation of enhancement strategies. Conservation Biology 7:289–301.

Haig, S. M., C. L. Gratto-Trevor, T. D. Mullins, and M. A. Colwell. 1997. Population identification of western hemisphere shorebirds throughout the annual cycle. Molecular Ecology 6:413–427.

Haig, S. M., T. R. Mace, and T. D. Mullins. 2003. Parentage and relatedness in the polyandrous Comb-crested Jacana using ISSRs. Journal of Heredity 94:302–309.

Haig, S. M., J. R. Walters, and J. H. Plissner. 1994b. Genetic evidence for monogamy in the Red-cockaded Woodpecker, a cooperative breeder. Behavioral Ecology and Sociobiology 23:295–303.

Hanski, I. 1992. Inferences from ecological incidence functions. American Naturalist 139:657–662.

Hanski, I. 1994a. Patch-occupancy dynamics in fragmented landscapes. Trends in Ecology and Evolution 9:131–135.

Hanski, I. 1994b. A practical model of metapopulation dynamics. Journal of Animal Ecology 63: 151–162.

Hanski, I. 1998. Metapopulation dynamics. Nature 396:41–49.

Hanski, I. 1999. Metapopulation Ecology. Oxford University Press, New York.

Hanski, I. 2002. Animal metapopulations in highly fragmented landscapes and PVA. Pages 86–108 *in* Population Viability Analysis (S. R. Beissinger and D. R. McCullough, Eds.). University of Chicago Press, Chicago.

Hanski, I., and O. Gaggiotti. 2004. Metapopulation biology: Past, present, and future. Pages 3–22 *in* Ecology, Genetics, and Evolution of Metapopulations (I. Hanski and O. Gaggiotti, Eds.). Elsevier Academic Press, New York.

Harcourt, A. H. 1995. Population viability estimates: Theory and practice for a wild gorilla population. Conservation Biology 9:134–142.

Harcourt, A. H. 2002. Empirical estimates of minimum viable population sizes for primates: Tens to tens of thousands? Animal Conservation 5:237–244.

Harris, R. B., L. A. Maguire, and M. L. Shaffer. 1987. Sample sizes for minimum viable population estimation. Conservation Biology 1: 72–76.

Harrison, S. 1994. Large-scale ecology and conservation biology: 35th symposium of the British Ecological Society, Southampton, England. Pages 111–128 *in* Metapopulations and Conservation (P. J. Edwards, R. M. May, and N. R. Webb, Eds.). Blackwell Scientific, Oxford, United Kingdom.

Harrison, S., and E. Bruna. 1999. Habitat fragmentation and large-scale conservation: What do we know for sure? Ecography 22:225–232.

Harrison, S., and C. Ray. 2002. Plant population viability and metapopulation-level processes. Pages 109–122 *in* Population Viability Analysis (S. R. Beissinger and D. R. McCullough, Eds.). University of Chicago Press, Chicago.

Hastings, A. 2003. Metapopulation persistence with age-dependent disturbance or succession. Science 301:1525–1526.

Hatten, J. R., and C. E. Paradzick. 2003. A multiscaled model of Southwestern Willow

Flycatcher breeding habitat. Journal of Wildlife Management 67:774–788.

Hedrick, P. W., and P. S. Miller. 1992. Conservation genetics: Techniques and fundamentals. Ecological Applications 2:30–46.

Heppell, S. S., J. R. Walters, and L. B. Crowder. 1994. Evaluating management alternatives for Red-cockaded Woodpeckers: A modeling approach. Journal of Wildlife Management 58: 479–487.

Hess, G. R. 1996a. Disease in metapopulation models: Implications for conservation. Ecology 77:1617–1632.

Hess, G. R. 1996b. Linking extinction to connectivity and habitat destruction in metapopulation models. American Naturalist 148:226–236.

Hilborn, R., and M. Mangel. 1997. The Ecological Detective: Confronting Models with Data. Princeton University Press, Princeton, New Jersey.

Hitchcock, C. L., and C. Gratto-Trevor. 1997. Diagnosing a shorebird local population decline with a stage-structured population model. Ecology 78:522–534.

Hoekman, S. T., L. S. Mills, D. W. Howerter, J. H. Devries, and I. J. Ball. 2002. Sensitivity analyses of the life cycle of midcontinental Mallards. Journal of Wildlife Management 66: 883–900.

Hof, J., and M. G. Raphael. 1997. Optimization of habitat placement: A case study of the Northern Spotted Owl in the Olympic Peninsula. Ecological Applications 7:1160–1169.

Hokit, D. G., B. M. Stith, and L. C. Branch. 1999. Effects of landscape structure in Florida scrub: A population perspective. Ecological Applications 9:124–134.

Hokit, D. G., B. M. Stith, and L. C. Branch. 2001. Comparison of two types of metapopulation models in real and artificial landscapes. Conservation Biology 15:1102–1113.

Holland, J. D., D. G. Bert, and L. Fahrig. 2004. Determining the spatial scale of species' response to habitat. BioScience 54:227–233.

Holmes, E. E. 2004. Beyond theory to application and evaluation: Diffusion approximations for population viability analysis. Ecological Applications 14:1272–1293.

Huston, M. A. 1992. Individual based forest succession models and the theory of plant competition. Pages 209–420 *in* Individual-based Models and Approaches in Ecology: Populations, Communities, and Ecosystems (D. L. DeAngelis and L. J. Gross, Eds.). Chapman and Hall, New York.

Ims, R. A., and N. G. Yoccoz. 1997. Studying transfer process in metapopulations: Emigration, migration, and colonization. Pages 247–265 *in* Metapopulation Biology: Ecology, Genetics, and Evolution (I. A. Hanski and M. E. Gilpin, Eds.). Academic Press, San Diego, California.

Inchausti, P., and H. Weimerskirch. 2001. Risks of decline and extinction of the endangered Amsterdam Albatross and the projected impact of long-line fisheries. Biological Conservation 100:377–386.

James, D. A., and J. C. Neal. 1986. Arkansas Birds. University of Arkansas Press, Fayetteville.

Jennings, M. D. 2000. Gap analysis: Concepts, methods, and recent results. Landscape Ecology 15:5–20.

Johnson, F. A., C. T. Moore, W. L. Kendall, J. A. Dubovsky, D. F. Caithamer, J. J. R. Kelley, and B. K. Williams. 1997. Uncertainty and the management of Mallard harvests. Journal of Wildlife Management 61:202–216.

Johnson, F. A., and B. K. Williams. 1999. Protocol and practice in the adaptive management of waterfowl harvests. Conservation Ecology 3: 1–14.

Johnson, J. B., and K. S. Omland. 2004. Model selection in ecology and evolution. Trends in Ecology and Evolution 19:101–108.

Jones, K. L. 2002. Refining the Whooping Crane studbook by incorporating microsatellite DNA and leg-banding analyses. Conservation Biology 16:789–799.

Jorgensen, S. E., and G. Bendoricchio. 2001. Fundamentals of Ecological Modeling, 3rd ed. Elsevier, Amsterdam, The Netherlands.

Kareiva, P., D. Skelly, and M. Ruckelshaus. 1997. Reevaluating the use of models to predict the consequences of habitat loss and fragmentation. Pages 156–166 *in* The Ecological Basis of Conservation: Heterogeneity, Ecosystems, and Biodiversity (S. T. A. Pickett, R. S. Ostfeld, M. Shachak, and G. E. Likens, Eds.). Chapman and Hall, New York.

Karl, J. W., P. J. Heglund, E. O. Garton, J. M. Scott, N. M. Wright, and R. L. Hutto. 2000. Sensitivity of species habitat-relationship model performance to factors of scale. Ecological Applications 10:1690–1705.

Kauffman, M. J., W. F. Frick, and J. Linthicum. 2003. Estimation of habitat-specific demography and population growth for Peregrine Falcons in California. Ecological Applications 13:1802–1816.

Keitt, B. S., C. Wilcox, B. R. Tershy, D. A. Croll, and C. J. Donlan. 2002. The effect of feral cats on the population viability of Black-vented Shearwaters (*Puffinus opisthomelas*) on Natividad Island, Mexico. Animal Conservation 5:217–223.

Kilgo, J. C., R. A. Sargent, B. R. Chapman, and K. V. Miller. 1998. Effect of stand width and adjacent habitat on breeding bird communities in bottomland hardwoods. Journal of Wildlife Management 62:72–83.

KINDVALL, O. 2000. Comparative precision of three spatially realistic simulation models of metapopulation dynamics. Ecological Bulletin 48: 101–110.

KING, A. W., L. K. MANN, W. W. HARGROVE, T. L. ASHWOOD, AND V. H. DALE. 2000. Assessing the persistence of an avian population in a managed landscape: A case study with Henslow's Sparrow at Fort Knox, Kentucky. Oak Ridge National Laboratory, Oak Ridge, Tennessee.

KREMENTZ, D. G., AND J. S. CHRISTIE. 1999. Scrub-successional bird community dynamics in young and mature longleaf pine–wiregrass savannahs. Journal of Wildlife Management 63:803–814.

LACY, R. C. 2004. SPARKS—Single Population Analysis Record Keeping System. Chicago Zoological Society, Chicago.

LACY, R. C., M. BORBET, AND J. P. POLLAK. 2005. VORTEX: A stochastic simulation of the extinction process, version 9.5. Chicago Zoological Society, Brookfield, Illinois.

LACY, R. C., N. R. FLESNESS, AND U. S. SEAL. 1989. Puerto Rican Parrot (*Amazona vittata*) population viability analysis and recommendations. Captive Breeding Specialist Group, Apple Valley, Minnesota.

LAMBERSON, R. H., R. MCKELVEY, B. R. NOON, AND C. VOSS. 1992. A dynamic analysis of Northern Spotted Owl viability in a fragmented forest landscape. Conservation Biology 6:505–512.

LAMBERSON, R. H., B. R. NOON, C. VOSS, AND K. S. MCKELVEY. 1994. Reserve design for territorial species: The effects of patch size and spacing on the viability of the Northern Spotted Owl. Conservation Biology 8:185–195.

LA MONTAGNE, J. M., R. L. IRVINE, AND E. E. CRONE. 2002. Spatial patterns of population regulation in sage grouse (*Centrocercus* spp.) population viability analysis. Journal of Animal Ecology 71:672–682.

LANDE, R. 1988a. Genetics and demography in biological conservation. Science 241:1455–1460.

LANDE, R. 1988b. Demographic models of the Northern Spotted Owl (*Strix occidentalis caurina*). Oecologia 75:601–607.

LANDE, R. 1993. Risks of population extinction from demographic and environmental stochasticity and random catastrophes. American Naturalist 142:911–927.

LANDE, R. 1995. Mutation and conservation. Conservation Biology 9:782–791.

LANDE, R. 2002. Incorporating stochasticity in population viability analysis Pages 18–40 *in* Population Viability Analysis (S. R. Beissinger and D. R. McCullough, Eds.). University of Chicago Press, Chicago.

LEBRETON, J.-D., K. P. BURNHAM, J. CLOBERT, AND D. R. ANDERSON. 1992. Modeling survival and testing biological hypotheses using marked animals: A unified approach. Ecological Monographs 62:67–118.

LEFKOVITCH, L. P. 1965. The study of population growth in organisms grouped by stages. Biometrics 21:1–18.

LESLIE, P. H. 1945. On the use of matrices in certain population mathematics. Biometrika 35: 183–212.

LESLIE, P. H. 1948. Some further notes on the use of matrices in population mathematics. Biometrika 35:213–245.

LESLIE, P. H. 1966. The intrinsic rate of increase and the overlap of successive generations in a population of guillemots (*Uria aalge* Pont.). Journal of Animal Ecology 35:291–301.

LETCHER, B. H., J. A. PRIDDY, J. R. WALTERS, AND L. B. CROWDER. 1998. An individual-based, spatially-explicit simulation model of the population dynamics of the endangered Red-cockaded Woodpecker. Biological Conservation 86:1–14.

LEVINS, R. 1969. Some demographic and genetic consequences of environmental heterogeneity for biological control. Bulletin of the Entomological Society of America 15:237–240.

LINDENMAYER, D. B., M. A. BURGMAN, H. R. AKÇAKAYA, R. C. LACY, AND H. P. POSSINGHAM. 1995. A review of the generic computer programs ALEX, RAMAS/space and VORTEX for modelling the viability of wildlife metapopulation. Ecological Modelling 82:161–174.

LINDENMAYER, D. B., AND H. P. POSSINGHAM. 1996. Ranking conservation and timber management options for Leadbeater's possum in southeastern Australia using population viability analysis. Conservation Biology 10:235–251.

LINDENMAYER, D. B., H. P. POSSINGHAM, R. C. LACY, M. A. MCCARTHY, AND M. L. POPE. 2003. How accurate are population models? Lessons from landscape-scale tests in a fragmented system. Ecology Letters 6:41–47.

LINK, W. A., AND P. F. DOHERTY. 2002. Scaling in sensitivity analysis. Ecology 83:3299–3305.

LIU, J. 1993. ECOLECON: An ECOLogical-ECONomic model for species conservation in complex forest landscapes. Ecological Modelling 70:63–87.

LIU, J., J. B. DUNNING, JR., AND H. R. PULLIAM. 1995. Potential effects of a forest management plan on Bachman's Sparrows (*Aimophila aestivalis*): Linking a spatially explicit model with GIS. Conservation Biology 9:62–75.

LOCKWOOD, J. L., M. S. ROSS, AND J. P. SAH. 2003. Smoke on the water: The interplay of fire and water flow on Everglades restoration. Frontiers in Ecology and the Environment 1:462–468.

Loiselle, B. A., C. A. Howell, C. Graham, J. M. Goerck, T. Brooks, K. G. Smith, and P. H. Williams. 2003. Avoiding pitfalls of using species-distribution models in conservation planning. Conservation Biology 17:1591–1600.

Lotts, K. C., T. A. Waite, and J. A. Vucetich. 2004. Reliability of absolute and relative predictions of population persistence based on time series. Conservation Biology 18:1224–1232.

Ludwig, D. 1996. Uncertainty and the assessment of extinction probabilities. Ecological Applications 6:1067–1076.

Ludwig, D. 1999. Is it meaningful to estimate a probability of extinction? Ecology 80:298–310.

Ludwig, D., and C. J. Walters. 2002. Fitting population viability into adaptive management. Pages 511–520 *in* Population Viability Analysis (S. R. Beissinger and D. R. McCullough, Eds.). University of Chicago Press, Chicago.

MacCluer, J. W., J. L. Vandeberg, B. Read, and O. A. Ryder. 1986. Pedigree analysis by computer simulation. Zoo Biology 5:147–160.

Mace, G. 1986. Genedrop: Computer Software for Genedrop Analyses. Zoological Society of London, London.

Mace, G. M., and R. Lande. 1991. Assessing extinction threats: Toward a re-evaluation of IUCN threatened species categories. Conservation Biology 5:148–157.

MacKenzie, D. I., J. D. Nichols, J. E. Hines, M. G. Knutson, and A. B. Franklin. 2003. Estimating site occupancy, colonization, and local extinction when a species is detected imperfectly. Ecology 84:2200–2207.

MacKenzie, D. I., J. D. Nichols, G. B. Lachman, S. Droege, J. A. Royle, and C. A. Langtimm. 2002. Estimating site occupancy rates when detection probabilities are less than one. Ecology 83: 2248–2255.

MacKenzie, D. I., J. D. Nichols, J. A. Royle, K. H. Pollock, L. L. Bailey, and J. E. Hines. 2006. Occupancy Estimation and Modeling. Academic Press, San Diego, California.

Maguire, L. A., G. F. Wilhere, and Q. Dong. 1995. Population viability analysis for Red-cockaded Woodpeckers in the Georgia Piedmont. Journal of Wildlife Management 59:533–542.

Mangel, M., and C. Tier. 1994. Four facts every conservation biologist should know about persistence. Ecology 75:607–614.

Maynard Smith, J. 1974. Models in Ecology. Cambridge University Press, Cambridge, United Kingdom.

Maynard Smith, J. 1982. Evolution and the Theory of Games. Cambridge University Press, Cambridge, United Kingdom.

McCarthy, M. A., S. J. Andelman, and H. P. Possingham. 2003. Reliability of relative predictions in population viability analysis. Conservation Biology 17:982–989.

McCarthy, M. A., D. B. Lindenmayer, and H. P. Possingham. 2000. Testing spatial PVA models of Australian treecreepers (Aves: Climacteridae) in fragmented forest. Ecological Applications 10:1722–1731.

McCarthy, M. A., H. P. Possingham, J. R. Day, and A. J. Tyre. 2001. Testing the accuracy of population viability analysis. Conservation Biology 15:1030–1038.

McDonald, D. B., and H. Caswell. 1993. Matrix methods for avian demography. Pages 139–185 *in* Current Ornithology, vol. 10 (D. M. Power, Ed.). Plenum Press, New York.

McGarigal, K., and W. C. McComb. 1995. Relationships between landscape structure and breeding birds in the Oregon Coast Range. Ecological Monographs 65:235–260.

McGrath, M. T., S. DeStefano, R. A. Riggs, L. L. Irwin, and G. J. Roloff. 2003. Spatially explicit influences on Northern Goshawk nesting habitat in the interior Pacific Northwest. Wildlife Monographs, no. 154.

McKelvey, D., B. R. Noon, and R. H. Lamberson. 1993. Conservation planning for species occupying fragmented landscapes: The case of the Northern Spotted Owl. Pages 424–452 *in* Biotic Interactions and Global Change (P. M. Kareiva, J. G. Kingslover, and R. B. Huey, Eds.). Sinauer Associates, Sunderland, Massachusetts.

Meretsky, V. J., N. F. R. Snyder, S. R. Beissinger, D. A. Clendenen, and J. W. Wiley. 2000. Demography of the California Condor: Implications for reestablishment. Conservation Biology 14:957–967.

Mertz, D. B. 1971. The mathematical demography of the California Condor population. American Naturalist 105:437–453.

Miller, J. R., M. G. Turner, E. A. H. Smithwick, C. L. Dent, and E. H. Stanley. 2004. Spatial extrapolation: The science of predicting ecological patterns and processes. BioScience 54: 310–320.

Miller, R. S., and D. B. Botkin. 1974. Endangered species models and predictions. American Scientist 62:172–181.

Mills, L. S., D. F. Doak, and M. J. Wisdom. 1999. Reliability of conservation actions based on elasticity analysis of matrix models. Conservation Biology 13:815–829.

Mills, L. S., S. G. Hayes, C. Baldwin, M. J. Wisdom, J. Citta, D. J. Mattson, and K. Murphy. 1996. Factors leading to different viability predictions for a grizzly bear data set. Conservation Biology 10:863–873.

Mills, L. S., and M. S. Lindberg. 2002. Sensitivity analysis to evaluate the consequences of conservation actions. Pages 338–366 *in* Population Viability Analysis (S. R. Beissinger and D. R. McCullough, Eds.). University of Chicago Press, Chicago.

Mills, L. S., and P. E. Smouse. 1994. Demographic consequences of inbreeding in remnant populations. American Naturalist 144:412–431.

Mitchell, M. S., R. A. Lancia, and J. A. Gerwin. 2001. Using landscape-level data to predict the distribution of birds on a managed forest: Effects of scale. Ecological Applications 11: 1692–1708.

Moilanen, A. 1999. Patch occupancy models of metapopulation dynamics: Efficient parameter estimation using implicit statistical inference. Ecology 80:1031–1043.

Moilanen, A. 2004. SPOMSIM: Software for stochastic patch occupancy models of metapopulation dynamics. Ecological Modelling 179:533–550.

Mooij, W. M., R. E. Bennetts, W. M. Kitchens, and D. L. DeAngelis. 2002. Exploring the effect of drought extent and interval on the Florida Snail Kite: Interplay between spatial and temporal scales. Ecological Modelling 149:25–39.

Mooij, W. M., and D. L. DeAngelis. 1999. Error propagation in spatially explicit population models: A reassessment. Conservation Biology 13:930–933.

Mooij, W. M., and D. L. DeAngelis. 2003. Uncertainty in spatially explicit animal dispersal models. Ecological Applications 13: 794–805.

Morris, W. F., and D. F. Doak. 2002. Quantitative Conservation Biology: Theory and Practice of Population Viability Analysis. Sinauer Associates, Sunderland, Massachusetts.

Morrison, M. L., B. G. Marcot, and R. M. Mannan. 1992. Wildlife–Habitat Relationships: Concepts and Applications. University of Wisconsin Press, Madison.

Murdoch, W. W., E. McCauley, R. M. Nisbet, S. C. Gurney, and A. M. De Roos. 1992. Individual-based models: Combining testability and generality. Pages 18–35 *in* Individual-Based Models and Approaches in Ecology: Populations, Communities, and Ecosystems (D. L. DeAngelis and L. J. Gross, Eds.). Chapman and Hall, New York.

National Research Council (NRC). 1996. The Scientific Basis for the Preservation of the Mariana Crow. National Academy of Sciences, Washington, D.C.

Nichols, J. D. 1992. Capture–recapture models. BioScience 42:94–102.

Nichols, J. D. 2001. Using models in the conduct of science and management of natural resources. Pages 11–34 *in* Modeling in Natural Resource Management (T. M. Shenk and A. B. Franklin, Eds.). Island Press, Washington, D.C.

Nichols, J. D., G. L. Hensler, and P. W. Sykes, Jr. 1980. Demography of the Everglade Kite: Implications for population management. Ecological Modelling 9:215–232.

Nichols, J. D., and J. E. Hines. 2002. Approaches for the direct estimation of lambda, and demographic contributions to lambda, using capture–recapture data. Journal of Applied Statistics 29:539–568.

Nichols, J. D., J. E. Hines, J.-D. Lebreton, and R. Pradel. 2000. Estimation of contributions of a reverse-time capture–recapture approach. Ecology 81:3362–3376.

Nichols, J. D., J. E. Hines, K. H. Pollock, R. L. Hinz, and W. A. Link. 1994. Estimating breeding proportions and testing hypotheses about costs of reproduction with capture–recapture data. Ecology 75:2052–2065.

Nichols, J. D., F. A. Johnson, and B. K. Williams. 1995. Managing North American waterfowl in the face of uncertainty. Annual Review of Ecology and Systematics 26:177–199.

Nix, H. A. 1986. A biogeographic analysis of Australian elapid snakes Pages 4–15 *in* Atlas of Australian Elapid Snakes (R. Longmore, Ed.). CSIRO Publishing, Collingwood, Victoria, Australia.

Nolan, V., Jr. 1978. The ecology and behavior of the Prairie Warbler, *Dendroica discolor*. Ornithological Monographs, no. 26.

Noon, B. R., and K. S. McKelvey. 1996a. Management of the Spotted Owl: A case history in conservation biology. Annual Review of Ecology and Systematics 27:135–162.

Noon, B. R., and K. S. McKelvey. 1996b. A common framework for conservation planning: Linking individual and metapopulation models. Pages 139–165 *in* Metapopulations and Wildlife Conservation (D. R. McCullough, Ed.). Island Press, Washington, D.C.

Noon, B. R., and D. D. Murphy. 1997. Management of the Spotted Owl: The interaction of science, policy, politics, and litigation. Pages 432–441 *in* Principles of Conservation Biology (G. Meffe and R. Carroll, Eds.). Sinauer Associates, Sunderland, Massachusetts.

Noon, B. R., and J. R. Sauer. 1992. Population models for passerine birds: Structure, parameterization, and analysis. Pages 441–464 *in* Wildlife 2001: Populations (D. R. McCullough and R. H. Barrett, Eds.). Elsevier Applied Science, New York.

Norris, K. 2004. Managing threatened species: The ecological toolbox, evolutionary theory and declining-population paradigm. Journal of Applied Ecology 41:413–426.

Norris, K., and N. McCulloch. 2003. Demographic models and the management of endangered species: A case study of the critically endangered Seychelles Magpie Robin. Journal of Applied Ecology 40:890–899.

O'Connor, R. J. 2002. The conceptual basis of species distribution modeling: Time for a paradigm shift? Pages 25–33 *in* Predicting Species Occurrences: Issues of Accuracy and Scale (J. M. Scott, P. J. Heglund, M. L. Morrison, J. B. Haufler, M. G. Raphael, W. A. Wall, and F. B. Samson, Eds.). Island Press, Washington, D.C.

O'Connor, R. J., and M. T. Jones. 1997. Using hierarchical models to assess the ecological health of the nation. Transactions of the North American Wildlife and Natural Resources Conference 62:501–508.

O'Connor, R. J., and T. L. Wagner. 2004. A test of a regression-tree model of species distribution. Auk 121:604–609.

Olson, G. S., E. M. Glenn, R. G. Anthony, E. D. Forsman, J. A. Reid, P. J. Loschl, and W. J. Ripple. 2004. Modeling demographic performance of Northern Spotted Owls relative to forest habitat in Oregon. Journal of Wildlife Management 68:1039–1053.

Oro, D., J. S. Aguilar, J. M. Igual, and M. Louzao. 2004. Modeling demography and extinction risk in the endangered Balearic Shearwater. Biological Conservation 116:93–102.

Parysow, P., and D. J. Tazik. 2002. Assessing the effect of estimation error on population viability analysis: An example using the Black-capped Vireo. Ecological Modelling 155: 217–229.

Pascual, M. A., P. Kareiva, and R. Hilborn. 1997. The influence of model structure on conclusions about the viability and harvesting of Serengeti wildebeest. Conservation Biology 11:966–976.

Pearson, S. M. 1993. The spatial extent and relative influence of landscape-level factors on wintering bird populations. Landscape Ecology 8:3–18.

Peery, M. Z., B. H. Becker, and S. R. Beissinger. 2006a. Age ratios as estimators of productivity: Testing assumptions on an endangered seabird, the Marbled Murrelet. Auk 123: in press.

Peery, M. Z., B. H. Becker, and S. R. Beissinger. 2006b. Combining demographic and count-based approaches to identify source–sink dynamics: An example using an endangered seabird. Ecological Applications. In press.

Peery, M. Z., S. R. Beissinger, S. H. Newman, E. B. Burkett, and T. D. Williams. 2004. Applying the declining population paradigm: Diagnosing causes of poor reproduction in the Marbled Murrelet. Conservation Biology 18: 1088–1098.

Pereira, J. M. C., and R. M. Itami. 1991. GIS-based habitat modeling using logistic multiple regression: A study of the Mt. Graham red squirrel. Photogrammetric Engineering and Remote Sensing 57:1475–1486.

Peterson, A. T. 2003a. Projected climate change effects on Rocky Mountain and Great Plains birds: Generalities of biodiversity consequences. Global Change Biology 9:647–655.

Peterson, A. T. 2003b. Predicting the geography of species' invasions via ecological niche modeling. Quarterly Review of Biology 78:419–433.

Peterson, A. T. 2005. Kansas Gap Analysis: The importance of validating distributional models before using them. Southwestern Naturalist 50:230–236.

Peterson, A. T., S. L. Egbert, V. Sánchez-Cordero, and K. P. Price. 2000. Geographic analysis of conservation priority: Endemic birds and mammals in Veracruz, Mexico. Biological Conservation 93:85–94.

Peterson, A. T., and R. D. Holt. 2003. Niche differentiation in Mexican birds: Using point occurrences to detect ecological innovation. Ecology Letters 6:774–782.

Peterson, A. T., and D. A. Kluza. 2003. New distributional modelling approaches for gap analysis. Animal Conservation 6:47–54.

Peterson, A. T., M. A. Ortega-Huerta, J. Bartley, V. Sánchez-Cordero, J. Soberón, R. H. Buddemeier, and D. R. B. Stockwell. 2002a. Future projections for Mexican faunas under global climate change scenarios. Nature 416: 626–629.

Peterson, A. T., D. R. B. Stockwell, and D. A. Kluza. 2002b. Distributional prediction based on ecological niche modeling of primary occurrence data. Pages 617–623 *in* Predicting Species Occurrences: Issues of Accuracy and Scale (J. M. Scott, P. J. Heglund, M. L. Morrison, J. B. Haufler, M. G. Raphael, W. A. Wall, and F. B. Samson, Eds.). Island Press, Washington, D.C.

Peterson, A. T., and D. A. Vieglais. 2001. Predicting species invasions using ecological niche modeling: New approaches from bioinformatics attack a pressing problem. BioScience 51:363–371.

Pfister, C. A. 1998. Patterns of variance in stage-structured populations: Evolutionary predictions and ecological implications. Proceedings of the National Academy of Sciences USA 95: 213–218.

Pfister, C. A., and M. Wang. 2005. Beyond size: Matrix projection models for populations where size is an incomplete descriptor. Ecology 86:2673–2683.

Pielou, E. C. 1981. The usefulness of ecological models: A stock-taking. Quarterly Review of Biology 56:17–31.

Pimm, S. L. 1991. The Balance of Nature? University of Chicago Press, Chicago.

Pimm, S. L., and O. L. Bass, Jr. 2002. Rangewide risks to large populations: The Cape Sable Sparrow as a case history. Pages 406–424 *in* Population Viability Analysis (S. R. Beissinger and D. R. McCullough, Eds.). University of Chicago Press, Chicago.

Pimm, S. L., and A. Redfearn. 1988. The variability of population densities. Nature 334:613–614.

Polansky, S., J. D. Camm, A. R. Solow, B. Csuti, D. White, and R. Ding. 2000. Choosing reserve networks with incomplete species information. Biological Conservation 94:1–10.

Possingham, H. P., S. J. Andelman, B. R. Noon, S. Trombulak, and H. R. Pulliam. 2001. Making smart conservation decisions. Pages 225–244 *in* Conservation Biology: Research Priorities for the Next Decade (M. E. Soulé and G. H. Orians, Eds.). Island Press, Washington, D.C.

Possingham, H. P., D. B. Lindenmayer, and G. N. Tuck. 2002. Decision theory for population viability analysis. Pages 470–489 *in* Population Viability Analysis (S. R. Beissinger and D. R. McCullough, Eds.). University of Chicago Press, Chicago.

Pradel, R. 1996. Utilization of capture–mark–recapture for the study of recruitment and population growth rate. Biometrics 52:703–709.

Price, M. V., and M. Gilpin. 1996. Modelers, mammalogists, and metapopulations: Designing Stephens' kangaroo rat reserves. Pages 217–240 *in* Metapopulations and Wildlife Conservation (D. R. McCullough, Ed.). Island Press, Washington, D.C.

Pulliam, H. R. 1988. Sources, sinks and population regulation. American Naturalist 135:652–661.

Pulliam, H. R., J. B. Dunning, Jr., and J. Liu. 1992. Population dynamics in complex landscapes: A case study. Ecological Applications 2:165–177.

Pulliam, H. R., J. Liu, J. B. Dunning, Jr., D. J. Stewart, and T. D. Bishop. 1995. Modeling animal populations in changing landscapes. Ibis 137 (Supplement):120–126.

Ralls, K., J. D. Ballou, and A. R. Templeton. 1988. Estimates of lethal equivalents and the cost of inbreeding in mammals. Conservation Biology 2:185–193.

Ralls, K., S. R. Beissinger, and J. F. Cochrane. 2002. Guidelines for using PVA in endangered species management. Pages 521–550 *in* Population Viability Analysis (S. R. Beissinger and D. R. McCullough, Eds.). University of Chicago Press, Chicago.

Ralls, K., and B. L. Taylor. 1997. How viable is population viability analysis? Pages 228–235 *in* The Ecological Basis of Conservation: Heterogeneity, Ecosystem, and Biodiversity (S. T. A. Pickett, R. S. Ostfeld, M. Shachak, and G. E. Likens, Eds.). Chapman and Hall, New York.

Reed, D. H., J. J. O'Grady, B. W. Brook, J. D. Ballou, and R. Frankham. 2003. Estimates of minimum viable population sizes for vertebrates and factors influencing those estimates. Biological Conservation 113:23–34.

Reed, J. M., P. D. Doerr, and J. R. Walters. 1988. Minimum viable population size of the Red-cockaded Woodpecker. Journal of Wildlife Management 52:385–391.

Reed, J. M., C. S. Elphick, and L. W. Oring. 1998. Life-history and viability analysis of the endangered Hawaiian Stilt. Biological Conservation 84:35–45.

Reed, J. M., L. S. Mills, J. B. Dunning, Jr., E. S. Menges, K. S. McKelvey, R. Frye, S. R. Beissinger, M. C. Anstett, and P. Miller. 2002. Emerging issues in population viability analysis. Conservation Biology 16:7–19.

Rice, J. A., T. J. Miller, L. B. Crowder, E. A. Marschall, A. S. Trebitz, and D. L. DeAngelis. 1993. Growth rate variation and larval survival: Inferences from an individual-based, size-dependent predation model. Canadian Journal of Fisheries and Aquatic Sciences 50:133–142.

Rodrigues, A. S. L., H. R. Akçakaya, S. J. Andelman, M. I. Bakarr, L. Boitani, T. M. Brooks, J. S. Chanson, L. D. C. Fishpool, G. A. B. da Fonseca, K. J. Gaston, and others. 2004. Global gap analysis: Priority regions for expanding the global protected-area network. BioScience 54:1092–1100.

Ruckelshaus, M., C. Hartway, and P. Kareiva. 1997. Assessing the data requirements of spatially explicit dispersal models. Conservation Biology 11:1298–1306.

Russello, M. A., and G. Amato. 2004. *Ex situ* population management in the absence of pedigree information. Molecular Ecology 13: 2829–2840.

Ryan, M. R., B. G. Root, and P. M. Mayer. 1993. Status of Piping Plovers in the Great Plains of North America: A demographic simulation model. Conservation Biology 7:581–585.

Saab, V. 1999. Importance of spatial scale to habitat use by breeding birds in riparian forests: A hierarchical analysis. Ecological Applications 9:135–151.

Sabo, J. L., E. E. Holmes, and P. Kareiva. 2004. Efficacy of simple viability models in ecological risk assessment: Does density dependence matter? Ecology 85:328–341.

Sæther, B.-E., and Ø. Bakke. 2000. Avian life history variation and contribution of demographic traits to the population growth rate. Ecology 81: 642–653.

Sæther, B.-E., and S. Engen. 2002. Including uncertainties in population viability analysis using population prediction intervals. Pages 191–212 *in* Population Viability Analysis (S. R. Beissinger and D. R. McCullough, Eds.). University of Chicago, Chicago.

Sæther, B.-E., S. Engen, A. P. Møller, M. E. Visser, E. Matthysen, W. Fiedler, M. M. Lambrechts, P. H. Becker, J. E. Brommer, J. Dickinson, and others. 2005. Time to extinction of bird populations. Ecology 86:693–700.

Sandercock, B. K., and S. R. Beissinger. 2002. Estimating rates of population change for a Neotropical parrot with ratio, mark–recapture and matrix methods. Journal of Applied Statistics 29:589–607.

Sandercock, B. K., K. Martin, and S. J. Hannon. 2005. Life history strategies in extreme environments: Comparative demography of Arctic and alpine ptarmigan. Ecology 86: 2176–2186.

Sauer, J. R., and N. A. Slade. 1987. Size-based demography of vertebrates. Annual Review of Ecology and Systematics 18:71–90.

Savignac, C., A. Desrochers, and J. Huot. 2000. Habitat use by Pileated Woodpeckers at two spatial scales in eastern Canada. Canadian Journal of Zoology 78:219–225.

Schemske, D. W., B. C. Husband, M. H. Ruckelshaus, C. Goodwillie, I. M. Parker, and J. G. Bishop. 1994. Evaluating approaches to the conservation of rare and endangered plants. Ecology 75:584–606.

Schiegg, K., J. R. Walters, and J. A. Priddy. 2005. Testing a spatially explicit, individual-based model of Red-cockaded Woodpecker population dynamics. Ecological Applications 15: 1495–1503.

Scott, J. M., F. Davis, B. Csuti, R. Noss, B. Butterfield, C. Groves, H. Anderson, S. Caicco, F. D'Erchia, T. C. Edwards, Jr., and others. 1993. Gap analysis: A geographic approach to protection of biological diversity. Wildlife Monographs, no. 123.

Scott, J. M., P. J. Heglund, M. L. Morrison, J. B. Haufler, M. G. Raphael, W. A. Wall, and F. B. Samson, Eds. 2002. Predicting Species Occurrences: Issues of Accuracy and Scale. Island Press, Washington, D.C.

Scott, J. M., T. H. Tear, and F. W. Davis, Eds. 1996. Gap Analysis: A Landscape Approach to Biodiversity Planning. American Society of Photogrammetry and Remote Sensing, Baltimore, Maryland.

Shaffer, M. L. 1981. Minimum population sizes for species conservation. BioScience 31:131–134.

Shaffer, M. L. 1983. Determining minimum viable population sizes for the grizzly bear. International Conference on Bear Research and Management 5:133–139.

Shaffer, M. L. 1987. Minimum viable populations: Coping with uncertainty. Pages 69–86 *in* Viable Populations for Conservation (M. E. Soulé, Ed.). Cambridge University Press, Cambridge, United Kingdom.

Shaffer, M. L., L. H. Watchman, W. J. Snape III, and I. K. Latchis. 2002. Population viability analysis and conservation policy. Pages 123–146 *in* Population Viability Analysis (S. R. Beissinger and D. R. McCullough, Eds.). University of Chicago Press, Chicago.

Shrader-Frechette, K. S., and E. D. McCoy. 1993. Method in Ecology: Strategies for Conservation. Cambridge University Press, Cambridge, United Kingdom.

Shugart, H. H. 1984. A Theory of Forest Dynamics: The Ecological Implications of Forest Succession Models. Springer-Verlag, New York.

Silvertown, J., M. Franco, and E. Menges. 1996. Interpretation of elasticity matrices as an aid to the management of plant populations for conservation. Conservation Biology 10:591–597.

Sjogren-Gulve, P., and I. Hanski. 2000. Metapopulation viability analysis using occupancy models. Ecological Bulletin 48:53–71.

Sjogren-Gulve, P., and C. Ray. 1996. Using logistic regression to model metapopulation dynamics: Large-scale forestry extirpates the pool frog. Pages 111–137 *in* Metapopulations and Wildlife Conservation (D. R. McCullough, Ed.). Island Press, Washington, D.C.

Sklar, F. H., M. J. Chimney, S. Newman, P. McCormick, D. Gawlik, S. L. Miao, C. McVoy, W. Said, J. Newman, C. Coronado, and others. 2005. The ecological–societal underpinnings of Everglades restoration. Frontiers in Ecology and the Environment 3:161–169.

Sklar, F. H., H. C. Fitz, Y. Wu, R. Van Zee, and C. McVoy. 2001. The design of ecological landscape models for Everglades restoration. Ecological Economics 37:379–401.

Smith, K. G., and D. G. Catanzaro. 1996. Predicting vertebrate distributions for gap analysis: Potential problems in constructing the models. Pages 163–170 *in* Gap Analysis: A Landscape Approach to Biodiversity Planning (J. M. Scott, T. H. Tear, and F. W. Davis, Eds.). American Society of Photogrammetry and Remote Sensing, Baltimore, Maryland.

Smith, K. G., W. F. Limp, R. S. Dzur, D. G. Catanzaro, and M. E. Garner. 1998. State Wide Biodiversity Mapping for Arkansas. Center for Advanced Spatial Technologies, University of Arkansas, Fayetteville.

Smith, M. R., P. W. Mattocks, and K. M. Cassidy. 1997. Breeding birds of Washington State: Location data and predicted distributions. Seattle Audubon Society Publications in Zoology, no. 1.

Soulé, M. E. 1980. Thresholds for survival: Maintaining fitness and evolutionary potential. Pages 151–169 *in* Conservation Biology: An Evolutionary–Ecological Approach (M. E. Soulé and B. A. Wilcox, Eds.). Sinauer Associates, Sunderland, Massachusetts.

Soulé, M. E., Ed. 1987. Viable Populations for Conservation. Cambridge University Press, New York.

Stacey, P. B., V. A. Johnson, and M. L. Taper. 1997. Migration within metapopulations: The impact upon local population dynamics. Pages 267–292 *in* Metapopulation Biology: Ecology, Genetics, and Evolution (I. A. Hanski, and M. E. Gilpin, Eds.). Academic Press, San Diego, California.

Stacey, P. B., and M. Taper. 1992. Environmental variation and the persistence of small populations. Ecological Applications 2:18–29.

Staples, D. F., M. L. Taper, and B. Dennis. 2004. Estimating population trend and process variation for PVA in the presence of sampling error. Ecology 85:923–929.

Stauffer, D. F. 2002. Linking populations and habitats: Where have we been? Where are we going? Pages 53–61 *in* Predicting Species Occurrences: Issues of Accuracy and Scale (J. M. Scott, P. J. Heglund, M. L. Morrison, J. B. Haufler, M. G. Raphael, W. A. Wall, and F. B. Samson, Eds.). Island Press, Washington, D.C.

Stearns, S. C. 1981. On measuring fluctuating environments: Predictability, constancy, and contingency. Ecology 62:184–199.

Stephens, D. W., and J. R. Krebs. 1986. Foraging Theory. Princeton University Press, Princeton, New Jersey.

Stephens, P. A., W. J. Sutherland, and R. P. Freckleton. 1999. What is the Allee effect? Oikos 87:185–190.

Stith, B. M., J. W. Fitzpatrick, G. E. Woolfenden, and B. Pranty. 1996. Classification and conservation of metapopulations: A case study of the Florida Scrub Jay. Pages 187–215 *in* Metapopulations and Wildlife Conservation (D. R. McCullough, Ed.). Island Press, Washington, D.C.

Stockwell, D. R. B. 1999. Genetic algorithms II. Pages 123–144 *in* Machine Learning Methods for Ecological Applications (A. H. Fielding, Ed.). Kluwer Academic Publishers, Boston.

Stockwell, D. R. B., and I. R. Noble. 1992. Induction of sets of rules from animal distribution data: A robust and informative method of analysis. Mathematics and Computers in Simulation 33:385–390.

Stockwell, D. R. B., and D. Peters. 1999. The GARP modeling system: Problems and solutions to automated spatial prediction. International Journal of Geographic Information Systems 13:143–158.

Stockwell, D. R. B., and A. T. Peterson. 2003. Comparison of resolution of methods used in mapping biodiversity patterns from point occurrence data. Ecological Indicators 3:213–221.

Taylor, B. L. 1995. The reliability of using population viability analysis for risk classification of species. Conservation Biology 9:551–558.

Taylor, B. L., P. R. Wade, U. Ramakrishnan, M. Gilpin, and H. R. Akçakaya. 2002. Incorporating uncertainty in population viability analyses for the purpose of classifying species by risk. Pages 239–252 *in* Population Viability Analysis (S. R. Beissinger and D. R. McCullough, Eds.). University of Chicago Press, Chicago.

ter Braak, J. F., I. A. Hanski, and J. Verboom. 1998. The incidence function approach to modeling of metapopulation dynamics. Pages 167–188 *in* Modeling Spatiotemporal Dynamics in Ecology (J. Bascompte and R. V. Sole, Eds.). Springer-Verlag, New York.

Thomas, C. D., A. Cameron, R. E. Green, M. Bakkenes, L. J. Beaumont, Y. C. Collingham, B. F. N. Erasmus, M. F. de Siqueira, A. Grainger, L. Hannah, and others. 2004. Extinction risk from climate change. Nature 427:145–148.

Thomas, C. D., and I. Hanski. 2004. Metapopulation dynamics in changing environments: Butterfly responses to habitat and climate change. Pages 489–514 *in* Ecology, Genetics, and Evolution of Metapopulations (I. Hanski and O. Gaggiotti, Eds.). Elsevier Academic Press, New York.

Thomas, J. W., E. D. Forsman, J. B. Lint, E. C. Meslow, B. R. Noon, and J. Verner. 1990. A conservation strategy for the Northern Spotted Owl. Report of the Interagency Scientific Committee to address the conservation of the Northern Spotted Owl. U.S. Department of Agriculture, Forest Service, U.S. Department of Interior, Bureau of Land Management, Fish and Wildlife Service, and National Park Service. U. S. Government Printing Office 791-171/20026, Washington, D.C.

Treves, A., L. Naughton-Treves, E. K. Harper, D. J. Mladenoff, R. A. Rose, T. A. Sickley, and A. P. Wydeven. 2004. Predicting human–carnivore conflict: A spatial model derived from 25 years of data on wolf predation on livestock. Conservation Biology 18:114–125.

Tuljapulkar, S. 1989. An uncertain life: Demography in random environments. Theoretical Population Biology 35:227–294.

URBAN, D. L. 2005. Modeling ecological processes across scales. Ecology 86:1996–2006.

U.S. FISH AND WILDLIFE SERVICE (USFW). 1985. Endangered species recovery plan for Red-cockaded Woodpeckers. U.S. Fish and Wildlife Service, Atlanta, Georgia.

VAN HORN, B. 2002. Approaches to habitat modeling: The tensions between pattern and process and between specificity and generality. Pages 63–72 *in* Predicting Species Occurrences: Issues of Accuracy and Scale (J. M. Scott, P. J. Heglund, M. L. Morrison, J. B. Haufler, M. G. Raphael, W. A. Wall, and F. B. Samson, Eds.). Island Press, Washington, D.C.

VAN TIENDEREN, P. H. 2000. Elasticities and the link between demographic and evolutionary dynamics. Ecology 81:666–679.

VILLARD, M.-A., K. E. FREEMARK, AND G. MERRIAM. 1992. Metapopulation theory and Neotropical migrant birds in temperate forests: An empirical investigation. Pages 474–482 *in* Ecology and Conservation of Neotropical Migrant Landbirds (J. M. Hagan III and D. W. Johnston, Eds.). Smithsonian Institution Press, Washington, D.C.

WADE, P. R. 2002. Bayesian population viability analysis. Pages 213–238 *in* Population Viability Analysis (S. R. Beissinger and D. R. McCullough, Eds.). University of Chicago Press, Chicago.

WALTERS, C. J. 1986. Adaptive Management of Natural Resources. McGraw-Hill, New York.

WALTERS, J. R., L. B. CROWDER, AND J. A. PRIDDY. 2002. Population viability analysis for Red-cockaded Woodpeckers using an individual-based model. Ecological Applications 12:249–260.

WANG, J. 2004. Monitoring and managing genetic variation in group breeding populations without individual pedigrees. Conservation Genetics 5:813–825.

WAPLES, R. S. 2002. Definition and estimation of effective population size in the conservation of endangered species. Pages 147–168 *in* Population Viability Analysis (S. R. Beissinger and D. R. McCullough, Eds.). University of Chicago Press, Chicago.

WEBSTER, M. S., P. P. MARRA, S. M. HAIG, S. BENSCH, AND R. T. HOLMES. 2002. Links between worlds: Unraveling migratory connectivity. Trends in Ecology and Evolution 17:76–83.

WENNERGREN, U., M. RUCKELHAUS, AND P. KAREIVA. 1995. The promise and limitations of spatial models in conservation biology. Oikos 74: 349–356.

WHITE, D., P. G. MINOTTI, M. J. BARCZAK, J. C. SIFNEOS, K. E. FREEMARK, M. V. SANTELMANN, C. F. STEINITZ, A. R. FIESTER, AND E. M. PRESTON. 1997. Assessing risks to biodiversity from future landscape change. Conservation Biology 11:349–360.

WHITE, G. W., A. B. FRANKLIN, AND T. M. SHENK. 2002. Estimating parameters of PVA models from data on marked animals. Pages 169–190 *in* Population Viability Analysis (S. R. Beissinger and D. R. McCullough, Eds.). University of Chicago Press, Chicago.

WIENS, J. A. 1997. The emerging role of patchiness in conservation biology. Pages 93–107 *in* The Ecological Basis of Conservation: Heterogeneity, Ecosystems, and Biodiversity (S. T. A. Pickett, R. S. Ostfeld, M. Shachak, and G. E. Likens, Eds.). Chapman and Hall, New York.

WIENS, J. A., B. VAN HORNE, AND B. R. NOON. 2002. Integrating landscape structure and scale into natural resource management. Pages 23–67 *in* Integrating Landscape Ecology into Natural Resource Management (J. Liu, and W. W. Taylor, Eds.). Cambridge University Press, New York.

WILLIAMS, B. K., J. D. NICHOLS, AND M. J. CONROY. 2002. Analysis and Management of Animal Populations: Modeling, Estimation, and Decision Making. Academic Press, San Diego, California.

WILLIAMS, S. E., E. E. BOLITHO, AND S. FOX. 2003. Climate change in Australian tropical rainforests: An impending environmental catastrophe. Proceedings of the Royal Society of London, Series B 270:1887–1892.

WISDOM, M. J., AND L. S. MILLS. 1997. Sensitivity analysis to guide population recovery: Prairie-chickens as an example. Journal of Wildlife Management 61:302–312.

WISDOM, M. J., L. S. MILLS, AND D. F. DOAK. 2000. Life-stage simulation analysis: Estimating vital rate effects on population growth for conservation. Ecology 81:628–641.

WITH, K. A. 1997. The application of neutral landscape models in conservation biology. Conservation Biology 11:1069–1080.

WITH, K. A. 2004. Metapopulation dynamics: Perspectives from landscape ecology. Pages 23-44 *in* Ecology, Genetics, and Evolution of Metapopulations (I. Hanski and O. Gaggiotti, Eds.). Elsevier Academic Press, New York.

WITH, K. A. 2005. Assessing extinction risk in Neotropical migratory songbirds. Pages 449–453 *in* Principles of Conservation Biology (M. J. Groom, G. K. Meffe, and C. R. Carroll, Eds.). Sinauer Associates, Sunderland, Massachusetts.

WITH, K. A., AND A. W. KING. 2001. Analysis of landscape sources and sinks: The effect of spatial pattern on avian demography. Biological Conservation 100:75–88.

WOOTON, J. T., AND D. A. BELL. 1992. A metapopulation model of the Peregrine Falcon in California: Viability and management strategies. Ecological Applications 2:307–321.

WRIGHT, S. 1931. Evolution in Mendelian populations. Genetics 16:97–159.